Master Math: Algebra

By
Debra Anne Ross

NO LONGER PROPERTY OF
SEATTLE PUBLIC LIBRARY

Course Technology PTR
A part of Cengage Learning

Australia, Brazil, Japan, Korea, Mexico, Singapore, Spain, United Kingdom, United States

COURSE TECHNOLOGY
CENGAGE Learning™

Master Math: Algebra
Debra Anne Ross

Publisher and General Manager, Course Technology PTR: Stacy L. Hiquet

Associate Director of Marketing: Sarah Panella

Manager of Editorial Services: Heather Talbot

Marketing Manager: Jordan Casey

Senior Acquisitions Editor: Emi Smith

Interior Layout: Shawn Morningstar

Cover Designer: Jeff Cooper

Illustrations and Equations: Judith Littlefield

Indexer: Larry Sweazy

Proofreader: Jenny Davidson

Printed in Canada
2 3 4 5 6 7 11 10 09

© 1996, 2010 Debra Anne Ross

ALL RIGHTS RESERVED. No part of this work covered by the copyright herein may be reproduced, transmitted, stored, or used in any form or by any means graphic, electronic, or mechanical, including but not limited to photocopying, recording, scanning, digitizing, taping, Web distribution, information networks, or information storage and retrieval systems, except as permitted under Section 107 or 108 of the 1976 United States Copyright Act, without the prior written permission of the publisher.

For product information and technology assistance, contact us at **Cengage Learning Customer and Sales Support, 1-800-354-9706.**

For permission to use material from this text or product, submit all requests online at **cengage.com/permissions.**

Further permissions questions can be e-mailed to **permissionrequest@cengage.com.**

Library of Congress Control Number: 2009924537
ISBN-13: 978-1-59863-980-3
ISBN-10: 1-59863-980-3

Course Technology, a part of Cengage Learning
20 Channel Center Street
Boston, MA 02210
USA

Cengage Learning is a leading provider of customized learning solutions with office locations around the globe, including Singapore, the United Kingdom, Australia, Mexico, Brazil, and Japan. Locate your local office at: **international.cengage.com/region.**

Cengage Learning products are represented in Canada by Nelson Education, Ltd.

For your lifelong learning solutions, visit **courseptr.com.**
Visit our corporate Web site at **cengage.com.**

Table of Contents

Acknowledgments

I sincerely thank Dr. Melanie McNeil, Professor of Chemical Engineering at San Jose State University, for reading this book for accuracy and for all her helpful comments. I am grateful to Dr. Channing Robertson, Professor of Chemical Engineering at Stanford University, for reviewing this book and, in general, for his sagacious guidance. I especially thank my mother, Maggie Ross, for reading this book and for her editorial help.

Without my wonderful agent, Sidney B. Kramer, and the staff of Mews Books, the *Master Math* series would not have been published. Thank you, Sidney! I am also thankful to Ron Fry and the staff of Career Press for their work in publishing and launching the original *Master Math* books as a successful series.

I am grateful to Emi Smith, Senior Acquisitions Editor, and Course Technology, a part of Cengage Learning, for invigorating the *Master Math* series and improving the presentation. I especially thank Shawn Morningstar's proficient work on layout and illustrations. I also very much appreciate Judith Littlefield's work on illustrations and equations. Much thanks to Jenny Davidson for proofreading, Jeff Cooper for cover design, Larry Sweazy for indexing, as well as Stacy L. Hiquet, Sarah Panella, Heather Talbot, and Jordan Casey.

Finally, I deeply appreciate my beautiful and brilliant husband, David A. Lawrence, who worked side-by-side with me as we meticulously edited text and figures.

About the Author

Debra Anne Ross Lawrence is the author of six books of the *Master Math* series: *Basic Math and Pre-Algebra*, *Algebra*, *Pre-Calculus*, *Calculus*, *Trigonometry*, and *Geometry*. She earned a double Bachelor of Arts degree in biology and chemistry with honors from the University of California at Santa Cruz and a Master of Science degree in chemical engineering from Stanford University.

Her research experience encompasses investigating the photosynthetic light reactions using a dye laser, studying the eye lens of diabetic patients, creating a computer simulation program of physiological responses to sensory and chemical disturbances, genetically engineering bacteria cells for over-expression of a protein, and designing and fabricating biological reactors for in-vivo study of microbial metabolism using nuclear magnetic resonance spectroscopy.

Debra was a member of a small team of scientists and engineers who developed and brought to market the first commercial biosensor system. She managed an engineering group responsible for scale-up of combinatorial synthesis for pharmaceutical development.

She also managed intellectual property for a scientific research and development company. Debra's work has been published in scientific journals and/or patented.

Debra is also the author of *The 3:00 PM Secret: Live Slim and Strong Live Your Dreams* and *The 3:00 PM Secret 10-Day Dream Diet*. She is the coauthor with her husband, David A. Lawrence, of *Arrows Through Time: A Time Travel Tale of Adventure, Courage, and Faith*. Debra is President of GlacierDog Publishing and Founder of GlacierDog.com. When Debra is not engaged in all-season mountaineering near her Alaska home, she is endeavoring to understand the incomprehensible workings of the universe.

Introduction

Algebra is the second book in the *Master Math* series. The series also includes *Basic Math*, *Pre-Calculus*, *Geometry*, *Trigonometry*, and *Calculus*. The *Master Math* series presents the general principles of mathematics from grade school through college including arithmetic, algebra, geometry, trigonometry, pre-calculus, and introductory calculus.

Algebra is a comprehensive algebra book that explains the subject matter in a way that makes sense to the reader. It begins with the most basic principles and progresses through more advanced topics to prepare a student for pre-calculus and calculus. *Algebra* explains the principles and operations of algebra, provides step-by-step procedures and solutions and presents examples and applications.

Algebra is a reference book for middle school and high school students that explains and clarifies the algebra principles they are learning in school. It is also a comprehensive reference source for students currently learning pre-calculus and calculus. *Algebra* is invaluable for students, parents, tutors, and anyone needing a comprehensive algebra reference source.

The information provided in each book and in the series as a whole is progressive in difficulty and builds on itself, which allows the reader to gain perspective on the connected nature of mathematics.

The skills required to understand every topic presented are explained in an earlier chapter or book within the series. Each of the books contains a complete table of contents and a comprehensive index, so that specific subjects, principles, and formulas can be easily found. The books are written in a simple style that facilitates understanding and easy referencing of sought-after principles, definitions, and explanations.

Algebra and the *Master Math* series are not replacements for textbooks but rather reference and teaching books providing explanations and perspective. The *Master Math* series would have been invaluable to me during my entire education from grade school through graduate school. There is no other source that provides the breadth and depth of the *Master Math* series in a single book or series.

Finally, mathematics is a language—the universal language. A person struggling with mathematics should approach it in the same fashion he or she would approach learning any other language. If someone moves to a foreign country, he or she does not expect to know the language automatically. It takes practice and contact with a language in order to master it. After a short time in the foreign country he or she would not say, "I do not know this language well yet. I must not have an aptitude for it." Yet many people have this attitude toward mathematics. If time is spent learning and practicing the principles, mathematics will become familiar and understandable. Don't give up.

Chapter 1

Translating Problems into Algebraic Equations

1.1 Introduction to Algebra

• Algebra is an extension of arithmetic and is used to solve problems. If a situation can be described in terms of an equation, then the equation can be solved for the unknown quantity and an answer can be obtained. Algebraic equations are created to enable the determination of some quantity.

• For example, suppose you need $20.00 to buy your mother some flowers, and you know you can sell lemonade for $1.00 per glass. How many glasses must you sell? An algebraic equation can be designed to answer this question.

Let's call the number of glasses that must be sold, "X." How many glasses at $1.00 per glass must be sold to make $20.00?

In other words, you must sell X glasses at $1.00 per glass to make $20.00. Therefore, X multiplied by $1.00 per glass will equal $20.00.

In equation form, this is written:

$$(X) \times (\$1.00 \text{ per glass}) = (\$20.00)$$

This simple algebraic equation provides a means to determine how many glasses must be sold. Solving for X involves algebraic techniques.

1.2 Translating English into Algebraic Equations

• Certain words are used to indicate certain mathematical operations. It is useful to identify these words in a word problem and use them to create algebraic equations.

• Addition may be represented by the following: *added to, plus, sum, in all, altogether, increased by, the sum of, more, total.*

• Subtraction may be represented by the following: *difference, how many less, subtracted from, how many more, reduced by, decreased by, minus, less, take away.*

• Multiplication may be represented by the following: *multiplied by, of, times, product, multiply, twice* (multiply by 2), *double* (multiply by 2).

• Division may be represented by the following: *divided by, quotient, divides, ratio.*

• Raise to a power may be represented by: *squared* (raise to the 2nd power), *cubed* (raise to the 3rd power).

• Set equal to may be represented by the following: *equals, is equal, is, are, were, was, the sum is, the difference is, the quotient is, the product is, the same as, results in, the result is, makes, leaves, yields, gives.*

• Indication to use a letter to represent a number is represented by the following: *an unknown, a variable, a number, what.*

• The following are examples of word phrases translated into algrebra:

38 is 8 more than 3 times a number.
Choose x to represent the unknown number.
The equation is $3x + 8 = 38$.

The square of a number doubled less 5 results in 11.
Choose y to represent the unknown number.
The equation is $(2y)^2 - 5 = 11$.

The sum of 5 and the cube of a number is 13.
Choose n to represent the unknown number.
The equation is $5 + n^3 = 13$.

1.3 Algebra Terminology

• Algebra terminology consists of certain symbols and words that are used to represent certain identities and definitions. The following is a list of many of these words and symbols:

• *Letters* are used to represent numbers. A letter can be represented by a, b, c, A, B, C, x, y, z, etc.

• A *variable* is a letter that represents one or more numbers. Letters representing unknown numbers are often called variables because the unknown may involve more than one number. Also, if the circumstance an equation is describing changes, the unknown may change. A variable can be represented by x, y, z, a, b, c, d, m, n, and other letters including

Greek letters. Note that, generally, letters at the end of the alphabet are used to represent variables, and letters at the beginning of the alphabet are used to represent constants.

- Often there are numbers in a problem that will not change; these are called *constants*. Constants are generally known quantities and are represented by the number itself or by a letter.

- If a variable and a constant are multiplied together, the constant is called a *coefficient*. The sign of the constant is included as part of the coefficient. For example, in 3y, the 3 is the coefficient and y is the variable, and in −4x, the −4 is the coefficient and x is the variable.

- When letters (representing a number) are multiplied with each other, or a letter is multiplied with a number, the multiplication sign may not be displayed.

$$3n = (3)(n) = 3 \times n$$

$$yn = (y)(n) = y \times n$$

- When numbers are multiplied with each other, the multiplication sign is displayed to prevent confusion.

$$2 \times 3 \neq 23$$

- As evident in the arithmetic section, a *numerical expression* is an expression involving arithmetic operations such as addition, subtraction, multiplication, and division. For example, $2 + 3$ and $5 \div 2$ are numerical expressions.

If variables are present in a numerical expression, the expression is called an *algebraic expression.* For example, $2x + 3$ and $x \div 4y$ are algebraic expressions.

- An *equation* is defined as a mathematical relationship or statement involving an equality where there is an equal sign present. An arithmetic equation is made up of arithmetic expressions and contains an equal sign. An algebraic equation includes algebraic expressions and contains an equal sign.

- In an algebraic equation, the numbers in between the $+$, $-$, or $=$ signs that are multiplied or divided with each other are called *terms.* The sign of the term is determined by the + or – sign preceding it, and the sign is considered part of the term. If there is no sign before a term, the term is assumed positive. For example, in the expression $3x-2$, $3x$ is the first term and -2 is the second term. If $(3x-2)$ is grouped inside parentheses, it is considered one term. In the expression $(3x-2)+2x$, $(3x-2)$ is the first term and $2x$ is the second term.

This may seem confusing, but once an equation is simplified, the terms within that equation may change. For example, given the expression $3(3x-2)+2x$, using the distributive property (described in Sections 2.2 and 2.3), the 3 can be multiplied into each term in the parentheses resulting in $9x-6+2x$. The first expression had two terms, $3(3x-2)$ and $2x$, and the new expression has three terms, $9x$, 6, and $2x$.

• *Inequalities* are represented by the symbols for greater than and less than, and describe expressions in which the value of the expression on one side of the symbol is greater than the value of the expression on the other side of the symbol. The symbols are $>$, $<$, $\geq$, and $\leq$, representing greater than, less than, greater than or equal to, and less than or equal to, respectively.

• In arithmetic, when two or more numbers are multiplied together, the numbers are considered *factors* of the product. In algebra, when two or more numbers and variables are multiplied together, the numbers and/or variables are also considered factors of the product. For example, in $(18)(x) = 18x$, both 18 and x are factors; in $(z)(x) = zx$, both z and x are factors; and in x^2, both x and x are factors.

• *Polynomials* are algebraic expressions with each variable in the expression having a positive integer exponent. If there is no exponent shown with a variable, an invisible "1" exponent is present. Examples of polynomials are:

$4x^4 - 2x^2 + 4$

$-5x^2 + 3x + 9$

$-5x^2 - 8$

(x is a variable.)

(See Section 5.1 for an explanation of polynomials.)

• *Linear equations* are equations in which the variables do not have any exponents other than 1. These equations, if plotted, will produce a straight line. For example, $2x + 1 = 5$ is a linear equation. The general form of a linear equation is $Ax + By = C$, where A, B, and C are constants, and x and y are variables. Another general form of a linear equation is $y = mx + b$, where m is the slope of the line and b is where the line intercepts the y axis on a coordinate system. (See Chapter 9 on coordinate systems and graphing equations.)

• *Non-linear equations* are equations in which the variables have exponents with degrees other than 1. These equations, if plotted, will produce curved lines. For example, $2x^2 + 1 = 5$ is a non-linear equation.

1.4 Simple Word Problems

• This section contains examples of word problems and formulas involving perimeter, area, volume, price/profit, percent, distance, work, mixtures, and interest.

Important note: Techniques used to solve equations in this section are explained in Chapter 2, "Simplifying Algebraic Equations," and Chapter 3, "Solving Simple Algebraic Equations."

Perimeter Word Problems

• To solve perimeter problems, add the lengths of the sides. If two or more sides are equal, multiplication can be used. (See Chapter 1, "Geometry," in *Master Math: Pre-Calculus* or *Master Math: Geometry* for more details.)

• For example, if a triangle has sides with lengths 2 feet, 3 feet, and 4 feet, what is the perimeter?

$$2 \text{ feet} + 3 \text{ feet} + 4 \text{ feet} = 9 \text{ feet}$$

• For example, if a rectangle has sides with lengths 4 feet and 6 feet, what is the perimeter?

$$2(4\text{feet}) + 2(6 \text{ feet}) = 8 \text{ feet} + 12 \text{ feet} = 20 \text{ feet}$$

Area Word Problems

• To solve area problems that involve rectangles and squares, multiply length times width, or height times width, as appropriate. (See Chapter 1, "Geometry," in *Master Math: Pre-Calculus* or *Master Math: Geometry* for a complete discussion of area of planar objects and also surface area of three-dimensional objects.)

• The area of a rectangle is given by:

$$\text{Area} = (\text{Length})(\text{Width})$$

• The area of a triangle is given by:

Area = (1/2)(Base)(Height)

• For example, if a rectangle has a length of 3 feet and a width of 2 feet, what is its area?

(3 feet)(2 feet) = 6 feet^2

Volume Word Problems

• To solve volume problems that involve three-dimensional rectangular objects, multiply the length, width, and height of the object. (See Chapter 1, "Geometry," in *Master Math: Pre-Calculus* or *Master Math: Geometry* for a complete discussion of volume of three-dimensional objects.)

• The volume of a rectangular solid is given by:

Volume = (Length)(Width)(Height)

• For example, if a rectangular solid has a length of 3 feet, as width of 2 feet and a height of 4 feet, what is its volume?

(3 feet)(2 feet)(4 feet) = 24 feet^3

Price and Profit Word Problems

• Price is given by:

Total Price = (number of items)(cost per item)

- Sale price is given by:

  Sale Price = (original price) – (%discount)(original price)

- For example, if the sale price of \$24 for a chair represented a 25% discount off the original price, what was the original price of the chair?

  $\$24 = \$x - (0.25)(\$x)$

Where, x = original price.

  $\$24 = \$x(1 - 0.25)$

  $\$24 = \$x(0.75)$

  $\$24/(0.75) = \$x$

  $\$x = \$32$

  \$32 is the original price of the chair.

(See Chapters 2 and 3 for algebraic techniques used.)

- Profit is given by:

  Profit = sales price – cost

- For example, if a chair costs a store \$16, what price should the store owner charge his customers in order to make a gross profit of 50% of the cost of the chair?

Choose S to represent sales price:

Sales price – cost = profit

$\$S - \$16 = (50\%)(\$16)$

$\$S - \$16 = (0.5)(\$16)$

$\$S - \$16 = \$8$

$\$S = \$8 + \$16$

$\$S = \$24$

The sale price the owner should charge is $24.

Percent Word Problems

- To find what percent a number is of the total, multiply (number/total)(100).

- For example, Jeff got 17 of the 20 test problems correct. What percent correct did he get?

$$(17 \div 20)(100) = 85\%$$

- To find the amount of the total that a given percent represents, divide the percent by 100, then multiply by the total.

- For example, Laura spent 40% of the $350.00 she earned. How much did she have left?

First, find the amount she spent.

Change percent to number:

$(40\%)/(100) = 0.40$

Multiply by the total:

$(0.40)(\$350) = \$140$

Therefore, she spent $140.

Find the amount she has left:

$\$350 - \$140 = \$210$

Distance Word Problems

- Distance traveled is given by:

  Distance Traveled = (Rate)(Time)

- For example, a car travels an average speed of 60 miles per hour for 2 hours and then travels 30 miles per hour for 2 hours. How far did the car travel in 4 hours?

The car traveled 60 miles per hour for 2 hours, so it traveled:

(60 miles/hour)(2 hours) = 120 miles

Then the car traveled 30 miles per hour for 2 hours, so it traveled:

(30 miles/hour)(2 hours) = 60 miles

In 4 hours, the car traveled:

120 miles + 60 miles = 180 miles

• Note that when multiplying (miles/hour)(hours) = miles, the hours in the numerator cancel with the hours in the denominator resulting in miles. In general, knowing what the final units should be is helpful for verifying that the equation is set up properly.

Work Word Problems

• In work problems, the rates at which certain persons or machines work alone is often given, and the solution requires calculating the rate that they work together. Conversely, the rate they work together may be given and it is necessary to calculate the rates at which certain persons or machines work alone.

• The general formula for solving work problems is $1/n + 1/m = 1/h$, where n and m represent the number of hours it takes two people or machines to do a job when working alone, and h represents the number of hours it takes the two people or machines to do the job when they are working together. This relationship can be understood by considering that: In 1 hour person m does $1/m$ of the job, person n does $1/n$ of the job, and together they do $1/h$ of the job.

• For example, if Mary can plant 1,000 bulbs in 4 hours and Ned can plant 1,000 bulbs in 6 hours, how many hours will it take Mary and Ned to plant 1,000 bulbs if they work together?

$$1/4 + 1/6 = 1/h$$

A common denominator is 12. The equation can be written equivalently as:

$$3/12 + 2/12 = 1/h$$

$$5/12 = 1/h$$

Multiply both sides by h:

$$h(5/12) = 1$$

Divide both sides by 5/12:

$$h = 1/(5/12)$$

$$h = 12/5 = 2.4$$

It will take 2.4 hours when they are working together.

Mixture Word Problems

- To solve mixture problems, quantities and their proper units must be combined to determine the unknown result. Be careful to translate mixed units into the same units. For example, rather than adding feet and meters, translate all meters into feet (or vice versa) and then add the feet. Also, be careful to cancel the units as appropriate. For example, in the first example below, (pounds)($/pound)=($), where pounds canceled.

• Example: If you mix 6 pounds of cereal at a cost of \$1.20 per pound and 2 pounds of nuts at a cost of \$1.60 per pound, what is the cost per pound of the mixture?

Let X represent the cost/pound of the mixture.

(6 pounds)(\$1.20/pound) + (2 pounds)(\$1.60/pound) = (8 pounds)(\$X/pound)

After multiplying 6 with 1.20 and 2 with 1.60:

\$7.20 + \$3.20 = (8 pounds)(\$X/pound)

\$10.40 = (8 pounds)(\$X/pound)

Divide both sides of the equation by 8 pounds:

\$10.40/8 pounds = \$X/pound

\$1.30/pound = the cost of the mixture.

• Example: The chef has a recipe for butter cookies that requires a flour with 35% oat flour and the rest wheat flour. His assistant bought a 15-pound bag of flour that contains 20% oat flour in wheat flour. How much pure oat flour must the chef add to the 15-pound bag of flour to get a final mixture that has 35% oat flour?

Let y represent the number of pounds of pure oat flour the chef must add.

The chef has 15 pounds of 20% oat flour so his final mixture will be: 15 pounds of 20% + y pounds of 100% = (15 +y) pounds of 35%.

Rewriting percents as numbers (divide by 100):

$$(15 \text{ pounds})(0.20) + (y \text{ pounds})(1.00) = ((15 + y) \text{ pounds})(0.35)$$

Solve for y: (See Chapters 2 and 3 for techniques used.)

$$(15)(0.20) + y = (15 + y)(0.35)$$

$$3 + y = 5.25 + 0.35y$$

$$y - 0.35y = 5.25 - 3$$

$$0.65y = 2.25$$

$$y = 2.25/0.65$$

$$y = 3.46$$

Therefore, 3.46 pounds of oat flour must be added for a final mixture with 35% oat flour.

- Money and coin problems are mixture problems. In coin problems, consider the number and value of each type of coin, and the total value of all coins.

Interest Word Problems

- For *simple interest* problems remember:

  Interest earned = (Principal invested) × (Rate of interest) × (Time principal is invested)

  or $I = P \times R \times T$

• For example, if \$4,000 is invested at 3% simple annual interest, how much interest will be earned and how much will the investment be worth after 3 months?

For 1 year, 3% interest is earned, or 3/100 of the amount is earned in 12 months.

(0.03)(\$4,000)(1 year) = \$120 in 12 months

In 3 months:

(\$120)(3mo./12 mo.) = \$30 interest earned after 3 months

\$4,000 + \$30 = \$4,030 = value of investment after 3 months

• To calculate *compound interest*, first divide the amount of interest by the number of parts (time periods) that are compounded, so that if the interest is compounded quarterly, divide the interest by 4, if compounded semiannually, divide the interest by 2, etc. That fraction of interest will be applied to the investment amount each time period.

• For example, if \$5,000 is invested at 5% annual interest, compounded semiannually, what is the amount of interest earned and the value of the investment after 1 year?

At 6 months, half of the 5% or (2.5%) interest will be earned:

(\$5,000)(0.025) = \$125 is the interest earned in 6 months.

\$5,000 + (\$5,000)(0.025) = \$5,125 is the investment value in 6 months.

At 1 year, the second half of the 5% interest (or 2.5%) will be earned on the new value of the investment.

($5,125)(0.025) = $128.12 is the interest earned the next 6 months.

$5,125 + ($5,125)(0.025) = $5,253.12 is the investment value in 1 year.

- Note: The algebraic techniques used to solve the equations in this section are described in the following two chapters.

Chapter 2

Simplifying Algebraic Equations

2.1 Commutative, Associative, and Distributive Properties of Addition and Multiplication

• Addition and multiplication are said to be commutative and associative, and multiplication is said to be distributive over addition. The following equations are used to describe these properties.

• Addition and multiplication are said to be *commutative* because:

$$a + b = b + a$$

$$a \times b = b \times a$$

Where a and b each represent any real number.

• Addition and multiplication are said to be *associative* because:

$$(a + b) + c = a + (b + c)$$

$$(a \times b) \times c = a \times (b \times c)$$

Where a, b, and c each represent any real number.

• Multiplication is *distributive* over addition because:

$$a \times (b + c) = (a \times b) + (a \times c)$$

This can be equivalently written:

$$a(b + c) = ab + ac$$

Where a, b, and c each represent any real number.

2.2 Using Associative and Distributive Properties

• The associative property and the distributive property are often employed to make solving problems easier.

Using the Associative Property

• Numbers can be added or multiplied in any order, because changing the order will not change the resulting sum or product.

• **Example:**

$1 + 2 + 3 = 6$

$(1 + 2) + 3 = 6$

$1 + (2 + 3) = 6$

$(3 + 2) + 1 = 6$

$3 + (2 + 1) = 6$

$(2 + 3) + 1 = 6$

$2 + (3 + 1) = 6$

• **Example:**

$(1 + 2) + (3 + 4) = 10$

$(1 + 2 + 3) + 4 = 10$

$3 + (2 + 4) + 1 = 10$

$4 + (3 + 2 + 1) = 10$

$(4 + 3) + (2 + 1) = 10$

- **Example:**

$$2 \times 3 \times 5 = 30$$

$$3 \times 2 \times 5 = 30$$

$$5 \times 3 \times 2 = 30$$

$$5 \times 2 \times 3 = 30$$

- **Example:**

$$(1 + 2)(3 + 4) = 1\times3 + 1\times4 + 2\times3 + 2\times4$$
$$= 3 + 4 + 6 + 8 = 21$$

$$(3 + 4)(1 + 2) = 3\times1 + 3\times2 + 4\times1 + 4\times2$$
$$= 3 + 6 + 4 + 8 = 21$$

(Where each number in the first term was multiplied with each number in the second term. See Section 5.4, "Multiplication of Polynomials.")

Using the Distributive Property

- It is often convenient to use the distributive property when combinations of addition or subtraction with multiplication or division exist in a problem. Think of distributing the number that is outside the parentheses to the numbers inside. For example:

$$2(3 + 4) = (2)(3) + (2)(4) = 6 + 8 = 14$$

$$2(4 - 3) = (2)(4) - (2)(3) = 8 - 6 = 2$$

• In these examples, the numbers in the parentheses could have been added or subtracted first, but in more complex problems it is often necessary to use the distributive property to solve for unknown variables.

• The distributive property can be used on an expression to factor or unfactor. Factoring can be thought of as *undistributing* and distributing can be thought of as *unfactoring.*

Distributing: $2(3 + 5) = (2)(3) + (2)(5)$
2 is multiplied.

Factoring: $(2)(3) + (2)(5) = 2(3 + 5)$
2 is factored out.

Factored form: $2(3 + 5)$
This form has a common factor that is factored out of each term.

Unfactored form: $(2)(3) + (2)(5)$
The unfactored form has a common factor that is multiplied with each term in a set of parentheses.

2.3 Combining Like Terms in Algebraic Equations

• Identifying and combining like terms are important tools to use when solving algebraic equations.

• *Like terms* are two or more terms in an equation that contain the same variable factors raised to the same power such that only the coefficients can differ.

• The following are examples of the variable factors that are contained in terms. The variable factors must be the same in two or more terms in an equation in order for the terms to be considered like terms.

$a, b, c, d, x, y, z, a^2, b^2, c^3, d^4,$

$x^2, \sqrt{x}, \sqrt[3]{x}, ab, x^2b, ab^2,$

$abc, xy, x^2y, xyz, (a+b)$

• Coefficients are numbers that are often before a variable factor. The numbers representing the coefficients do not have to be the same in two or more terms in order for the terms to be considered like terms. For example, in the term $2x^2$, the 2 is the coefficient and the x^2 is the variable factor.

• Examples of like terms include:

$2a$ and $34a$

$36ab^2$ and ab^2

$2a$ and $2a$

$2(a+b)$ and $3(a+b)$

Note that the coefficients can be different.

- Examples of *not*-like terms include:

  $2a$ and $2b$

  $2a$ and $2ab$

  $\sqrt{x}$ and $\sqrt[3]{x}$

  $36ab$ and ab^2

  $2b$ and $2b^2$

Note that the variable factors are different.

- Remember that if no power or exponent is displayed, exponent "1" is implied. $a = a^1$

- Also remember, $x^{1/n} = \sqrt[n]{x}$

- Once like terms are identified in an equation, they can be combined according to their signs. Combining like terms is often necessary when solving algebraic equations. When like terms are combined, only the coefficients are added or subtracted.

- The following are examples of combining like terms:

  $2a + 34a = 36a$

  $36ab^2 - ab^2 = 35ab^2$

  $2a - 2a = 0$

$$2(a+b) + 3(a+b) = 5(a+b)$$

$$3\sqrt{2} + 6\sqrt{2} = 9\sqrt{2}$$

$$2a + 3b + 4a = 6a + 3b$$

$$36ab^2 - ab^2 + 36a^2b = 35ab^2 + 36a^2b$$

2.4 Simplifying Algebraic Equations by Removing Parentheses and Combining Like Terms

• Simplifying algebraic equations by removing parentheses and combining like terms are common techniques used when solving algebraic equations. The following examples describe the process of employing these techniques.

• **Example:** Simplify $3y - 2(y + 3) = 6$.

Use the distributive property to distribute (multiply) the -2 into the parentheses.

$$3y - 2y + (-2)(3) = 6$$

$$3y - 2y - 6 = 6$$

Combine like terms.

$$(3y - 2y) - 6 = 6$$

$$1y - 6 = 6$$

Isolate y by adding 6 to both sides of the equal sign.

$$y - 6 + 6 = 6 + 6$$

$$y = 12$$

(Remember, always treat both sides of the "=" sign equally.)

To check that the result is correct, substitute $y=12$ into the original equation.

$$3y - 2(y + 3) = 6$$

$$3(12) - 2(12 + 3) = 6$$

$$36 - 2(15) = 6$$

$$36 - 30 = 6$$

$$6=6$$

- **Example:** Simplify $6ab + 2(2ab - 3a) = 4$.

To remove the parentheses and combine like terms, use the distributive property to distribute (multiply) the 2 into the parentheses.

$$6ab + 2(2ab) - 2(3a) = 4$$

$$6ab + 4ab - 6a = 4$$

Combine like terms.

$$(6ab + 4ab) - 6a = 4$$

$$10ab - 6a = 4$$

(This equation cannot be further simplified.)

• In some equations, like terms cannot be combined until parentheses are removed. In some equations, removing the parentheses creates like terms.

• **Example:** Simplify $6ab + 2a(2b - 3) = 2$.

Remove the parentheses and combine like terms. Use the distributive property to distribute (multiply) the 2a into the parentheses (there are no like terms until the parentheses are removed).

$$6ab + 2a(2b) - 2a(3) = 2$$

$$6ab + 4ab - 6a = 2$$

Combine like terms.

$$(6ab + 4ab) - 6a = 2$$

$$10ab - 6a = 2$$

2.5 The General Order to Perform Operations in Algebra

• In general, when solving complicated equations that include terms with different types of variables, perform the operations using the following guidelines:

Perform the operations inside the parentheses beginning with the inner parentheses and reduce exponents and radicals, then perform multiplication, then division, then addition, and then subtraction. Remember if there is a number directly preceding a set of parentheses (not separated by a "+" or a "−" sign), it must be multiplied to each term inside the parentheses.

Chapter 3

Solving Simple Algebraic Equations

3.1 Solving Algebraic Equations That Have One Unknown Variable

• In this section, algebraic equations with one unknown variable are solved, including equations requiring one operation, equations requiring more than one operation, equations with like terms, equations with the unknown variable on both sides of the equal sign, and equations with parentheses.

• To solve an equation for the unknown variable, the variable must be isolated to one side of the equal sign so that:

Unknown variable = known numbers or values

• For example, if y is the unknown variable, y must be isolated to one side of the equal sign.

$y = 3 + 2$

$y = 5$

• In general, if a number or variable is added, subtracted, multiplied, or divided to or from one side of the equal sign, the same operation must be executed on the other side of the equal sign. Treat both sides of the equal sign equally.

• **Example:** Find the value of y in $y + 3 = 6$.

To isolate y, subtract 3 from both sides.

$y + 3 - 3 = 6 - 3$

$y + 0 = 3$

$y = 3$

Check the answer by substituting the result into the original equation, $y + 3 = 6$.

Substitute $y = 3$.

$3 + 3 = 6$

$6 = 6$

- **Example:** Find the value of z in $4z = 32$

To isolate z, divide both sides by 4.

$4z/4 = 32/4$

$z = 8$

Check the answer by substituting the result into the original equation, $4z = 32$.

Substitute $z = 8$.

$(4)(8) = 32$

$32 = 32$

- **Example:** Find the value of a in $a - 6 = 25$.

To isolate a, add 6 to both sides.

$a - 6 + 6 = 25 + 6$

$a + 0 = 31$

$a = 31$

Check the answer by substituting the result into the original equation, $a - 6 = 25$.

Substitute $a = 31$.

$31 - 6 = 25$

$25 = 25$

- **Example:** Find the value of x in $x/4 = 2$.

  To isolate x, multiply both sides by 4.

  $(x/4)(4) = (2)(4)$

  $(4x/4) = 8$

  $1x = 8$

  $x = 8$

Check the answer by substituting the result into the original equation, $x/4 = 2$.

Substitute $x = 8$.

$8/4 = 2$

$2 = 2$

- Each of these equations required one operation to isolate the unknown variable. To isolate the unknown variable in more complicated problems, first perform addition or subtraction of numbers or terms to isolate the term with the unknown variable, then perform multiplication or division to isolate the unknown variable. The following examples require more than one operation to isolate the unknown variable.

- **Example:** Find the value of x in $x/4 + 4 = 2$.

  To isolate x, first subtract 4 from both sides.

  $x/4 + 4 - 4 = 2 - 4$

  $x/4 + 0 = -2$

  $x/4 = -2$

  Multiply both sides by 4.

  $(x/4)(4) = (-2)(4)$

  $(4x/4) = -8$

  $1x = -8$

  $x = -8$

Check the answer by substituting into the original equation, $x/4 + 4 = 2$.

  $-8/4 + 4 = 2$

  $-2 + 4 = 2$

  $2 = 2$

- **Example:** Find the value of z in $-z - 4 = 2$.

  To isolate z, first add 4 to both sides.

  $-z - 4 + 4 = 2 + 4$

  $-z = 6$

  Divide both sides by -1.

(To isolate z, not $-z$, remember $z = 1z$ and $-z = -1z$.)

$$(-z)/(-1) = (6)/(-1)$$

Because $(-1)/(-1) = 1$,

$$z = -6$$

Check the answer by substituting into the original equation, $-z - 4 = 2$

$$-(-6) - 4 = 2$$

Remember, (negative) × (negative) = positive.

$$6 - 4 = 2$$

$$2 = 2$$

• To solve algebraic equations with like terms, first combine like terms, then isolate the unknown variable by performing addition or subtraction, followed by multiplication or division.

• **Example:** Find y in $8y + 5 - 3y = 10$.

Combine like terms by first rearranging.

$$8y - 3y + 5 = 10$$

Because $8y - 3y = (8-3)y = 5y$,

$$5y + 5 = 10$$

Subtract 5 from both sides.

$$5y + 5 - 5 = 10 - 5$$

$$5y = 5$$

Divide both sides by 5.

$$(5y)/5 = 5/5$$

$$y = 1$$

Check the answer by substituting into the original equation, $8y + 5 - 3y = 10$.

$$(8)(1) + 5 - (3)(1) = 10$$

$$8 + 5 - 3 = 10$$

$$13 - 3 = 10$$

$$10 = 10$$

- To solve algebraic equations where the unknown variable is on both sides of the equal sign, first combine like terms onto one side of the equal sign, then isolate the unknown variable by performing addition or subtraction, followed by multiplication or division.

- **Example:** Find x in $3x - 10 = -4 + x$.

Combine like terms by subtracting x from both sides.

$$3x - x - 10 = -4 + x - x$$

$$2x - 10 = -4$$

Add 10 to both sides.

$$2x - 10 + 10 = -4 + 10$$

$$2x = 6$$

Divide both sides by 2.

$$(2x)/2 = 6/2$$

$$x = 3$$

Check the answer by substituting into the original equation, $3x - 10 = -4 + x$.

$$3(3) - 10 = -4 + 3$$

$$9 - 10 = -4 + 3$$

$$-1 = -1$$

• To solve algebraic equations that have parentheses, first remove the parentheses using the distributive property, then combine like terms onto one side of the equal sign, and finally isolate the unknown variable by performing addition or subtraction, followed by multiplication or division.

• **Example:** Find x in $10 + 2(3x - 10) = -5(4 + x)$.

Remove parentheses using the distributive property.

$$10 + 6x - 20 = -20 - 5x$$

(Note that the negative sign was distributed.)

Combine like terms by adding 5x to both sides.

$$10 + 6x + 5x - 20 = -20 - 5x + 5x$$

$$10 + 11x - 20 = -20$$

$$10 - 20 + 11x = -20$$

$$-10 + 11x = -20$$

Add 10 to both sides.

$$-10 + 10 + 11x = -20 + 10$$

$$11x = -10$$

Divide both sides by 11.

$$(11x)/11 = -10/11$$

$$x = -10/11$$

Check the result by substituting into the original equation, $10 + 2(3x - 10) = -5(4 + x)$.

Substitute $x = -10/11$

$$10 + 2(3(-10/11) - 10) = -5(4 + (-10/11))$$

$$10 + 2((-30/11) - 10) = -5(4 + (-10/11))$$

Use distributive property to eliminate parentheses.

$$10 + -60/11 - 20 = -20 + 50/11$$

$$-60/11 - 10 = -20 + 50/11$$

Add 20 to both sides.

$$-60/11 - 10 + 20 = -20 + 20 + 50/11$$

$$-60/11 + 10 = 50/11$$

Add 60/11 to both sides.

$$-60/11 + 60/11 + 10 = 50/11 + 60/11$$

$$10 = 60/11 + 50/11$$

$$10 = 110/11$$

$$10 = 10$$

3.2 Solving Simple Algebraic Equations Containing Fractions

• Algebraic equations often contain fractions. The following techniques can be used to solve algebraic equations containing fractions.

• The first technique involves removing the fractions by multiplying each term by the lowest common denominator, then solving for the unknown.

• To remove the fractions and solve the equations:

1. First reduce each term using factoring.
2. Identify the lowest common denominator.
3. Multiply each term by the common denominator.
4. Solve the equation for the unknown variable using algebraic techniques.
5. Check the solution by substituting it into the original equation.

• **Example:** Solve 2x/3 − 2x/4 = 4 for x.

$$\frac{2x}{3} - \frac{2x}{4} = \frac{4}{1}$$

The second term can be reduced.

$$\frac{2x}{3} - \frac{1x}{2} = \frac{4}{1}$$

Identify the lowest common denominator.

Multiples of 3 are: 3, 6, 9, 12, ...

Multiples of 2 are: 2, 4, 6, 8, ...

6 is the lowest common multiple and lowest common denominator. Multiply each term by the common denominator.

$$\frac{(6)2x}{3} - \frac{(6)1x}{2} = \frac{(6)4}{1}$$

Cancel terms in each denominator with common factors in each numerator.

$$(2)2x - (3)1x = (6)4$$

Solve for the unknown variable.

$$4x - 3x = 24$$

$$1x = 24$$

$$x = 24$$

Check by substituting into the original equation.

$2(24)/3 - 2(24)/4 = 4$

$48/3 - 24/2 = 4/1$

The common denominator is 6.

Multiply each term by 6.

$$\frac{(6)48}{3} - \frac{(6)24}{2} = \frac{(6)4}{1}$$

Reduce each term.

$$\frac{(2)48}{1} - \frac{(3)24}{1} = \frac{(6)4}{1}$$

$96 - 72 = 24$

$24 = 24$

- **Example:** Solve $6/10 = 2x/(2x+4)$ for x.

Factor the denominator $(2x+4)$, then reduce each term.

$$\frac{6}{10} = \frac{2x}{2(x+2)}$$

$$\frac{3}{5} = \frac{x}{(x+2)}$$

The lowest common denominator that will contain each different factor in the two denominators is $5(x+2)$. Multiply each term by the common denominator.

$$\frac{3(5)(x+2)}{5} = \frac{x(5)(x+2)}{(x+2)}$$

Cancel terms in each numerator with common factors in its denominator to eliminate fractions. Then solve for the unknown variable.

$$3(x + 2) = x(5)$$

Use the distributive property.

$$3x + 6 = 5x$$

To combine like terms, subtract 3x from both sides.

$$6 = 5x - 3x$$

$$6 = 2x$$

Divide both sides by 2.

$$6/2 = x$$

$$3 = x$$

Check by substituting ($x = 3$) into the original equation.

$$\frac{6}{10} = \frac{2(3)}{2(3)+4}$$

$$\frac{6}{10} = \frac{6}{6+4}$$

$$\frac{6}{10} = \frac{6}{10}$$

- An alternative method used to solve problems involving two fractions, one on each side of the equal sign, is to cross-multiply the numerators and denominators of the opposite fractions.

$$\frac{6}{10} = \frac{2x}{2x + 4}$$

Cross multiply $(6)\times(2x+4)$ and set it equal to $(10)\times(2x)$.

$$(6)(2x+4) = (10)(2x)$$

Solve for the unknown variable.

Multiply and use the distributive property.

$$12x + 24 = 20x$$

Combine like terms. Subtract 12x from both sides.

$$24 = 20x - 12x$$

$$24 = 8x$$

Divide both sides by 8.

$$24/8 = x$$

$3 = x$ (This result agrees with the previous example.)

- Note that $\frac{a}{b} = \frac{x}{y}$ can be written as $ay = bx$.

3.3 Solving Simple Algebraic Equations Containing Radicals

- Algebraic equations often contain square roots, cubed roots, or fourth roots, etc. The following technique can be used when solving algebraic equations containing radicals.

- To solve equations that have square roots:

1. Isolate the radical term on one side of the equal sign and combine like terms.
2. Square each side of the equation to eliminate the radical sign.
3. Solve the equation for the unknown variable.
4. Check the solution by substituting into the original equation.

Note that if a cubed root or fourth root is involved, raise each side of the equation in Step 2 to the third or fourth power, rather than raising the second power.

- **Example:** Solve $2\sqrt{x-1} = 8$ for x.

Isolate the radical term on one side of the equal sign and combine like terms.

Divide both sides by 2.

$$\sqrt{x-1} = \frac{8}{2}$$

$$\sqrt{x-1} = 4$$

Square each side of the equation to eliminate the radical sign.

$(\sqrt{x - 1})^2 = 4^2$

$(\sqrt{x - 1})(\sqrt{x - 1}) = 16$

$\sqrt{(x - 1)(x - 1)} = 16$

$x - 1 = 16$

Add 1 to each side.

$x = 16 + 1$

$x = 17$

Check the answer by substituting into the original equation.

$2\sqrt{17 - 1} = 8$

$2\sqrt{16} = 8$

Divide both sides by 2.

$\sqrt{16} = \frac{8}{2}$

$\sqrt{16} = 4$

To remove the square root, square both sides.

$(\sqrt{16})^2 = 4^2$

$16 = 16$

(Note that $\sqrt{16} = 4$ because $4 \times 4 = 16$.)

- Remember,

$$(\sqrt{x})^2 = (x^{1/2})^2 = x$$

$$(\sqrt[3]{x})^3 = (x^{1/3})^3 = x$$

$$(\sqrt[4]{x})^4 = (x^{1/4})^4 = x$$

Chapter 4

Algebraic Inequalities

4.1 Solving Algebraic Inequalities with One Unknown Variable

• Instead of solving an algebraic equation with an equal sign, it may be necessary to solve an algebraic inequality. Inequalities are denoted by the symbols $>$, $<$, $\geq$, and $\leq$. (See the inequality section, 1.19, in *Basic Math and Pre-Algebra,* for a discussion of inequalities.)

• *Inequalities* are represented by the symbols for greater than ($>$) and less than ($<$), and greater than or equal to ($\geq$) and less than or equal to ($\leq$). Inequalities describe expressions in which the value of the expression on one side of the symbol is greater than or greater than or equal to the value of the expression on the other side of the symbol.

• When solving inequalities involving adding, subtracting, multiplying, and dividing positive and negative numbers, the following rules apply:

If a number is added to both sides of the inequality, the inequality sign remains unchanged.

If a number is subtracted from both sides of the inequality, the inequality sign remains unchanged.

If a positive number is multiplied to both sides of the inequality, the inequality sign remains unchanged.

If a positive number is divided into both sides of the inequality, the inequality sign remains unchanged.

If a negative number is multiplied to both sides of the inequality, the inequality sign reverses.

If a negative number is divided into both sides of the inequality, the inequality sign reverses.

• Except for adjusting the inequality sign as described in these rules, inequalities are solved using the same techniques that are used to solve equations.

- For example, solve for x in $3 - (x/3) \leq -5$.

Subtract 3 from both sides.

$$-(x/3) \leq -5 - 3$$

$$-(x/3) \leq -8$$

Multiply both sides by -3. (Sign reverses.)

$$x \geq (-8)(-3)$$

$$x \geq 24$$

Check the answer by substituting it into the original equation.

$$3 - (24/3) \leq -5$$

$$3 - 8 \leq -5$$

$$-5 \leq -5$$

- This example, $3 - (x/3) \leq -5$, can also be solved as follows:

Add 5 to both sides.

$$8 - (x/3) \leq 0$$

Add (x/3) to both sides.

$$8 \leq (x/3)$$

Multiply both sides by 3.

$$24 \leq x$$

Note that the answer is still the same. Also, substitute a value greater than 24 to check the direction of the sign.

Chapter 5

Polynomials

5.1 Definitions

• Polynomials are algebraic expressions that generally have more than one term.

• A *monomial* has one term. For example, $2x$, $-5x^2$, 8

• A *binomial* has two terms. For example, $2x + 8$, $3x - 5x^2$

• A *trinomial* has three terms. For example, $-5x^2 + 3x + 9$, $4x^4 - 2x^2 + 4x$

• The exponent of the variable in a term defines the *degree* of that term. For example, $2x^2$ has a degree of 2.

• The *degree* of a polynomial equation or expression is determined by the highest degree of the variable in any of its terms.

$4x^4 - 2x^2 + 4x$ has a degree of 4.

$5x^2 + 3x + 9$ has a degree of 2.

$-5x^2$ has a degree of 2.

$2x^2y$ has a degree of 2.

8 has a degree of 0.

(The degree of a constant is zero.)

$2x$ has a degree of 1.

(The degree of a variable without an exponent shown is one.)

• If the variable does not have an exponent, an invisible "1" is present. ($x^1 = x$.) Equations with terms having the degree of one are called linear equations because the graphs of these equations form straight lines. If at least one variable in one term has an exponent other than one, the equation is non-linear. Equations containing the degree of 2 (with exponents having positive integer values) are called quadratic equations and the graphs of these equations form curved lines.

5.2 Addition of Polynomials

• Addition of polynomials is performed by combining and adding like terms.

• **Example:** Add $(y^2 - 3y + 4)$ and $(y - 3y^2 + y^3)$.

$$(y^2 - 3y + 4) + (y - 3y^2 + y^3)$$

Remove parentheses and group like terms.

$$= y^3 + y^2 - 3y^2 - 3y + y + 4$$

Add like terms.

$$= y^3 - 2y^2 - 2y + 4$$

Therefore:

$$(y^2 - 3y + 4) + (y - 3y^2 + y^3) = y^3 - 2y^2 - 2y + 4$$

• **Example:** Add $2y(y^2 - 2y + 2)$ and $3(y - 4y^2 + 2y^3)$.

$2y(y^2 - 2y + 2) + 3(y - 4y^2 + 2y^3)$.

Multiply 2y and 3 into the parentheses and group like terms.

$= 2y^3 - 4y^2 + 4y + 3y - 12y^2 + 6y^3$

$= 2y^3 + 6y^3 - 4y^2 - 12y^2 + 4y + 3y$

Add like terms.

$= 8y^3 - 16y^2 + 7y$

Therefore:

$2y(y^2 - 2y + 2) + 3(y - 4y^2 + 2y^3) = 8y^3 - 16y^2 + 7y$

5.3 Subtraction of Polynomials

• To subtract polynomials, distribute the negative sign into the parentheses by multiplying it with each term, and combine like terms.

• **Example:** Subtract $(y^2 - 3y + 4)$ and $(y - 3y^2 + y^3)$.

Use parentheses as a reminder that the entire second polynomial is subtracted.

$(y^2 - 3y + 4) - (y - 3y^2 + y^3)$

Distribute the negative sign into parentheses by multiplying it with each term.

$= y^2 - 3y + 4 - y + 3y^2 - y^3$

Note that each sign of each term after the minus sign changes to the opposite sign.

Combine like terms.

$$= -y^3 + y^2 + 3y^2 - 3y - y + 4$$

$$= -y^3 + 4y^2 - 4y + 4$$

Therefore:

$$(y^2 - 3y + 4) - (y - 3y^2 + y^3) = -y^3 + 4y^2 - 4y + 4$$

5.4 Multiplication of Polynomials

• Multiplication of polynomials, including multiplying monomials, multiplying a monomial with a polynomial, multiplying two binomials, and multiplying polynomials with polynomials, is described in this section.

• To *multiply monomials,* multiply the numerical coefficients with each other and multiply the variables with each other.

• **Example:** Multiply the monomials:

$$(6x)(-3x^2) = (6)(-3)(x)(x^2) = -18x^3$$

• **Example:** Multiply the monomials:

$$(-5x^2y)^2 = (-5x^2y)(-5x^2y) = (-5)(-5)(x^2)(x^2)(y)(y)$$
$$= 25x^4y^2$$

• To *multiply a monomial with a polynomial*, use the distributive property to multiply the monomial with each term in the polynomial.

• **Example:** Multiply the monomial and polynomial:

$$6(4x + 8) = (6)(4x) + (6)(8) = 24x + 48$$

• **Example:** Multiply the monomial and the polynomial:

$$-10x(2x^2 - 6x - 2)$$

$$=(-10x)(2x^2) + (-10x)(-6x) + (-10x)(-2)$$

$$= -20x^3 + 60x^2 + 20x$$

• To *multiply two binomials*, use the distributive property to multiply each term in one binomial with each term in the other binomial. In other words, multiply the first term in the first binomial with each term in the second binomial, next multiply the second term in the first binomial with each term in the second binomial, then combine like terms.

• **Example:** Multiply the binomials:

$$(2x + 6)(3x + 4)$$

$$= (2x)(3x) + (2x)(4) + (6)(3x) + (6)(4)$$

$$= 6x^2 + 8x + 18x + 24$$

$$= 6x^2 + 26x + 24$$

Note that this technique is often called the FOIL method because the *First, Outer, Inner,* and *Last* terms are multiplied with each other.

• To *multiply polynomials with polynomials,* use the distributive property to multiply each term in the first polynomial with each term in the second polynomial. In other words, multiply the first term in the first polynomial with each term in the second polynomial, then multiply the second term in the first polynomial with each term in the second polynomial, then multiply the third term in the first polynomial with each term in the second polynomial, and so on. Finally, combine like terms.

• **Example:** Multiply the polynomials:

$$(2x - 4)(x^2 + 2x - 5)$$

$$= (2x)(x^2) + (2x)(2x) + (2x)(-5) + (-4)(x^2) + (-4)(2x) + (-4)(-5)$$

$$= 2x^3 + 4x^2 + -10x + -4x^2 + -8x + 20$$

$$= 2x^3 + 4x^2 + -4x^2 + -10x + -8x + 20$$

$$= 2x^3 - 18x + 20$$

5.5 Division of Polynomials

• Division of polynomials, including division of monomials, division of polynomials by monomials, and division of polynomials by polynomials, is described in this section.

• To *divide monomials,* write the division in the form of a fraction, then divide numerical coefficients with each other and divide like variable bases with each other.

• **Example:** Divide $(6x) \div (-3x^2)$:

$$\frac{6x}{-3x^2} = \frac{6x}{-3xx} = \frac{2}{-x} = -\frac{2}{x}$$

• To *divide polynomials by monomials*, use the distributive property and divide the monomial into each term of the polynomial, or multiply one-over the monomial with each term of the polynomial.

• **Example:** Divide $(4x + 8) \div 6$:

Multiply one-over the monomial with each term in the polynomial.

$$(4x + 8) \div 6 = (1/6)(4x + 8)$$

$$= (1/6)(4x) + (1/6)(8) = 4x/6 + 8/6 = \frac{2}{3}x + \frac{4}{3}$$

• **Example:** Divide $(2x^2 - 6x - 2) \div -2x$:

Multiply one-over the monomial with each term in the polynomial.

$$(2x^2 - 6x - 2) \div -2x = (-1/2x)(2x^2 - 6x - 2)$$

$$= (-1/2x)(2x^2) + (-1/2x)(-6x) + (-1/2x)(-2)$$

$$= \frac{-2x^2}{2x} + \frac{6x}{2x} + \frac{2}{2x} = -x + 3 + 1/x$$

• To *divide polynomials by polynomials*, use the long division format. To do this, divide the first term in the divisor into the first term in the dividend, then multiply the divisor by the first term in the quotient and write the product under the dividend. Next, subtract like terms and bring down the next term. Repeat this long division procedure until there are no more terms to bring down.

• **Example:** Divide $(-10x + 8 + 8x^2) \div (2x - 4)$:

Arrange the terms with the largest degrees first, and write in long division format.

$$\begin{array}{r|l} & \hfill ? \\ \hline 2x-4 & 8x^2-10x+8 \end{array}$$

Divide the first term in the divisor into the first term in the dividend, $8x^2 \div 2x = 4x$.

$$\begin{array}{r|l} & 4x \\ \hline 2x-4 & 8x^2-10x+8 \end{array}$$

Multiply the divisor with the first term in the quotient, $4x(2x-4) = 8x^2-16x$ and write the product under the dividend.

$$\begin{array}{r|l} & 4x \\ \hline 2x-4 & 8x^2-10x+8 \\ & \underline{8x^2-16x} \end{array}$$

Subtract like terms and bring down the next term.

$$\begin{array}{r|l} & 4x \\ \hline 2x-4 & 8x^2-10x+8 \\ & \underline{8x^2-16x} \\ & \hfill 6x+8 \end{array}$$

Find the second term in the quotient by dividing 2x into 6x, $6x \div 2x = 3$.

$$\begin{array}{r|l} & 4x + 3 \\ \hline 2x - 4 & 8x^2 - 10x + 8 \\ & \underline{8x^2 - 16x} \\ & \qquad\quad 6x + 8 \end{array}$$

Multiply the divisor with the second term in the quotient, $3(2x - 4) = 6x - 12$ and write the product under the dividend.

$$\begin{array}{r|l} & 4x + 3 \\ \hline 2x - 4 & 8x^2 - 10x + 8 \\ & \underline{8x^2 - 16x} \\ & \qquad\quad 6x + 8 \\ & \qquad\quad 6x - 12 \end{array}$$

Subtract like terms and, because there is no next term to bring down, write the remainder as a fraction of the difference over the divisor.

$$\begin{array}{r|l} & 4x + 3 \\ \hline 2x - 4 & 8x^2 - 10x + 8 \\ & \underline{8x^2 - 16x} \\ & \qquad\quad 6x + 8 \\ & \qquad\quad \underline{6x - 12} \\ & \qquad\qquad\quad 20 \end{array}$$

Therefore, $(-10x + 8 + 8x^2) \div (2x - 4) = 4x + 3 + \frac{20}{2x - 4}$

where, $\frac{20}{2x - 4} = \frac{20}{2(x - 2)} = \frac{10}{x - 2}$.

Therefore, the final answer is $4x + 3 + (10/(x-2))$.

- **Example:** Divide $(x^2 + 4) \div (x + 4)$:

Arrange the terms with the largest degrees first (fill in missing terms using zero), and write in long division format.

$$x + 4 \overline{\smash{\big)}\, x^2 + 0x + 4} \quad ?$$

Divide the first term in the divisor into the first term in the dividend, $x^2 \div x = x$.

$$\begin{array}{r} x \\ x + 4 \overline{\smash{\big)}\, x^2 + 0x + 4} \end{array}$$

Multiply the divisor with the first term in the quotient, $x(x + 4) = x^2 + 4x$ and write the product under the dividend.

$$\begin{array}{r} x \\ x + 4 \overline{\smash{\big)}\, x^2 + 0x + 4} \\ \underline{x^2 + 4x } \end{array}$$

Subtract like terms and bring down the next term.

$$\begin{array}{r} x \\ x + 4 \overline{\smash{\big)}\, x^2 + 0x + 4} \\ \underline{x^2 + 4x } \\ -4x + 4 \end{array}$$

Find the second term in the quotient by dividing x into $-4x$, $-4x \div x = -4$.

$$\begin{array}{r|l} & \;\; x - 4 \\ \cline{2-2} x + 4 & x^2 + 0x + 4 \\ & \underline{x^2 + 4x} \\ & \quad\; -4x + 4 \end{array}$$

Multiply the divisor with the second term in the quotient, $-4(x + 4) = -4x - 16$ and write the product under the dividend.

$$\begin{array}{r|l} & \;\; x - 4 \\ \cline{2-2} x + 4 & x^2 + 0x + 4 \\ & \underline{x^2 + 4x} \\ & \quad\; -4x + 4 \\ & \quad\; \underline{-4x - 16} \end{array}$$

Subtract like terms and, because there is no next term to bring down, write the remainder as a fraction of the difference over the divisor.

$$\begin{array}{r|l} & \;\; x - 4 \\ \cline{2-2} x + 4 & x^2 + 0x + 4 \\ & \underline{x^2 + 4x} \\ & \quad\; -4x + 4 \\ & \quad\; \underline{-4x - 16} \\ & \qquad\qquad 20 \end{array}$$

Therefore, $(x^2 + 4) \div (x + 4) = x - 4 + \dfrac{20}{x + 4}$.

5.6 Factoring Polynomials with a Common Monomial Factor

• When solving equations or simplifying expressions, it is often beneficial to factor polynomials containing a common monomial factor in each term.

• The factored form of a number is an expression of the number as a product of numbers that, when multiplied together, equal the number. Similarly, the factored form of a polynomial is an expression of the polynomial as a product of monomials and/or polynomials that, when multiplied together, equal the polynomial.

• When factoring a polynomial with a common monomial factor, factor out the greatest common factor.

• For example, factor the following:

$6 = (2)(3)$
Where 2 and 3 are both factors of 6.

$2a + 2b = 2(a + b)$
Where 2 is the greatest common factor.

$2x + 5x = x(2 + 5) = 7x$
Where x is the greatest common factor.

$2x^2 + 4x^2y = 2x^2(1 + 2y)$
Where $2x^2$ is the greatest common factor.

$12x^4 - 6x^3 + 3x^2 = 3x^2(4x^2 - 2x + 1)$
Where $3x^2$ is the greatest common factor.

5.7 Factoring Polynomial Expressions with the Form $ax^2 + bx + c$

• When solving equations or simplifying expressions containing trinomials in the form $ax^2 + bx + c$, it is generally beneficial to factor the trinomial.

• Factoring a trinomial in the form $ax^2 + bx + c$ results in two binomials. Factoring a trinomial is the reverse of multiplying two binomials.

• Recall the steps involved in multiplying binomials.

Multiply binomials $(x + 2)(x + 3)$.

$$(x + 2)(x + 3) = x^2 + 3x + 2x + (2)(3)$$

$$= x^2 + (2+3)x + (2)(3) = x^2 + 5x + 6$$

Multiply binomials $(4x + 2)(5x + 3)$.

$$(4x + 2)(5x + 3)$$

$$= (4x)(5x) + (4x)(3) + (2)(5x) + (2)(3)$$

$$= 20x^2 + 12x + 10x + 6 = 20x^2 + 22x + 6$$

Compare the factored binomial form with the trinomial form.

• If m and n represent numbers, compare the factored binomial form and the trinomial form of this simple expression:

$$(x + m)(x + n) = x^2 + nx + mx + mn \quad \text{(factored form)}$$

$$= x^2 + (m+n)x + mn \quad \text{(trinomial form)}$$

The coefficient $(m+n)$ is the sum of the numbers in the last term of each binomial. The trinomial has the form, $ax^2 + bx + c$, where $a = 1$, $b = (m+n)$, and $c = mn$.

• If p, q, m, and n represent numbers, compare the factored binomial form and the trinomial form of this more complicated expression:

$$(px + m)(qx + n) = pqx^2 + pnx + qmx + mn$$
(factored form)

$$= pqx^2 + (pn+qm)x + mn$$
(trinomial form)

The coefficient $(pn+qm)$ results from the sum of the products of the first and last terms of each binomial. The sum of the outer product $(px)(n)$ and the inner product $(m)(qx)$ is $(pn+qm)x$. The trinomial has the form, $ax^2 + bx + c$, where $a = qp$, $b = (qm+pn)$, and $c = mn$.

- To factor a trinomial:

1. Write the factored format ( )( ).
2. Find sets of two values that, when multiplied together, equal the first term of the trinomial.
3. Find sets of two values that, when multiplied together, equal the last term of the trinomial.
4. Choose the above sets such that the sum of the outer product and the inner product of the binomial is equal to the second term (the first degree term) of the trinomial. (Remember to be careful of negative signs.)
5. Multiply the resulting binomials to check that the original trinomial is obtained.

- **Example:** Factor the trinomial $x^2 + 5x + 6$.

Write the factored format ( )( ).

Find sets of two values that, when multiplied together, equal the first term of the trinomial.

The only set is: x and x.

$$(x \quad)(x \quad)$$

Find sets of two values that, when multiplied together, equal the last term of the trinomial.

The possible sets are 2 and 3 or 1 and 6.

Set 1: (2)(3)

Set 2: (1)(6)

Choose these sets such that the sum of the outer product and the inner product of the binomial is equal to the second term of the trinomial.

The second term of the trinomial is 5x.

Therefore: Outer product + inner product must equal 5x.

Set 1: $3x + 2x = 5x$

Set 2: $6x + -1x = 5x$

Because there are no negative signs in the original trinomial, Set 2 is eliminated.

Therefore, the Set 1 binomial, $(x + 2)(x + 3)$, must be the factored binomial.

Multiply the chosen binomial set to check that it produces the original trinomial.

$$(x + 2)(x + 3) = x^2 + 3x + 2x + 6 = x^2 + 5x + 6$$

Therefore, the factored form of $x^2 + 5x + 6$ is

$(x + 2)(x + 3)$.

- **Example:** Factor the trinomial $x^2 + 3x - 18$.

Write the factored format ( )( ).

Find sets of two values that, when multiplied, equal the first term of the trinomial.

The only set is x and x.

$(x\quad)(x\quad)$

Find sets of two values that, when multiplied, equal the last term of the trinomial.

The possible sets are 6 and −3, −6 and 3, 9 and −2, −9 and 2, 18 and −1, −18 and 1.

$(\ +6)(\ -3)$

$(\ -6)(\ +3)$

$(\ +9)(\ -2)$

$(\ -9)(\ +2)$

$(\ +18)(\ -1)$

$(\ -18)(\ +1)$

Choose these sets such that the sum of the outer product and the inner product of the binomial is equal to the second term of the trinomial. The second term of the trinomial is 3x.

Therefore: Outer product + inner product must equal 3x.

$(x + 6)(x - 3)$: outer product $= -3x$, inner product $= 6x$, sum $= 3x$

$(x - 6)(x + 3)$: outer product $= 3x$, inner product $= -6x$, sum $= -3x$

$(x + 9)(x - 2)$: outer product $= -2x$, inner product $= 9x$, sum $= 7x$

$(x - 9)(x + 2)$: outer product $= 2x$, inner product $= -9x$, sum $= -7x$

$(x + 18)(x - 1)$: outer product $= -1x$, inner product $= 18x$, sum $= 17x$

$(x - 18)(x + 1)$: outer product $= 1x$, inner product $= -18x$, sum $= -17x$

The set that has the sum of the outer product and inner product equal to 3x is $(x + 6)(x - 3)$. Therefore this set must be the factored binomial.

Multiply the chosen binomial set to check that it produces the original trinomial.

$$(x + 6)(x - 3) = x^2 - 3x + 6x - 18 = x^2 + 3x - 18$$

Therefore, the factored form of $x^2 + 3x - 18$ is $(x + 6)(x - 3)$.

- **Example:** Factor the trinomial $20x^2 + 22x + 6$.

Write the factored format ( )( ).

Find sets of two values that, when multiplied, equal the first term of the trinomial. The possible sets are 2x and 10x, 20x and 1x, 5x and 4x.

(2x)(10x)

(20x)(1x)

(5x)(4x)

Find sets of two values that, when multiplied, equal the last term of the trinomial. The possible sets are 2 and 3, 1 and 6.

(2)(3)

(1)(6)

Choose these sets such that the sum of the outer product and the inner product of the binomial is equal to the second term of the trinomial.

The second term of the trinomials is 22x.

Therefore, the sum of the outer product and the inner product must equal 22x. From the above possible sets:

(2x)(10x)

(20x)(1x)

(5x)(4x)

(2)(3)

(1)(6)

Make all possible combinations of the inner and outer products.

$(2x + 2)(10x + 3)$: outer product = $6x$, inner product = $20x$, sum = $26x$

$(20x + 2)(1x + 3)$: outer product = $60x$, inner product = $2x$, sum = $62x$

$(5x + 2)(4x + 3)$: outer product = $15x$, inner product = $8x$, sum = $23x$

$(2x + 1)(10x + 6)$: outer product = $12x$, inner product = $10x$, sum = $22x$

$(20x + 1)(1x + 6)$: outer product = $120x$, inner product = $1x$, sum = $121x$

$(5x + 1)(4x + 6)$: outer product = $30x$, inner product = $4x$, sum = $34x$

Because there are no negative signs in the original trinomial, negative numbers are not included. The set that has the sum of the outer product and the inner product equal to $22x$ is $(2x + 1)(10x + 6)$; therefore, this set must be the factored binomial.

Multiply the chosen binomial set to check that it produces the original trinomial.

$$(2x + 1)(10x + 6) = 20x^2 + 12x + 10x + 6$$
$$= 20x^2 + 22x + 6$$

Therefore, the factored form of $20x^2 + 22x + 6$ is

$(2x + 1)(10x + 6)$.

Special Binomial Products to Remember

• The *difference of two squares*, $x^2 - y^2$, factors to $(x + y)(x - y)$ because the sum of the inner and outer products will always equal zero. For example:

$$(x + y)(x - y) = x^2 - xy + xy - y^2 = x^2 - y^2$$

$$(x + 3)(x - 3) = x^2 - 3x + 3x - 3^2 = x^2 - 3^2 = x^2 - 9$$

• **Example:** Factor the difference of two squares $4x^2 - 25$.

What squared equals $4x^2$?

$$(2x)(2x) = 4x^2$$

What squared equals 25?

$$(5)(5) = 25$$

Therefore, $4x^2 - 25 = (2x + 5)(2x - 5)$.

Check by multiplying the resulting binomials.

$$(2x + 5)(2x - 5) = 4x^2 - 10x + 10x - 25$$

$$= 4x^2 + (10 - 10)x - 25 = 4x^2 + 0x - 25 = 4x^2 - 25$$

• The *sum of two squares*, $x^2 + y^2$, cannot be further factored.

• The *binomial squared* has the following form:

$(x + y)^2 = x^2 + 2xy + y^2$ because,

$$(x + y)(x + y) = x^2 + xy + xy + y^2 = x^2 + 2xy + y^2$$

$(x - y)^2 = x^2 - 2xy + y^2$ because,

$(x - y)(x - y) = x^2 - xy - xy + y^2 = x^2 - 2xy + y^2$

- **Example:** Factor $x^4 - 81$.

Using the difference of two squares (discussed previously):

$x^4 - 81 = (x^2 + 9)(x^2 - 9)$

Also, $(x^2 - 9)$ can be further factored using the difference of two squares:

$x^2 - 9 = (x + 3)(x - 3)$

Therefore, the factored form of $x^4 - 81$ is

$(x^2 + 9)(x + 3)(x - 3)$

To check, multiply $(x^2 + 9)(x + 3)(x - 3)$.

First multiply $(x + 3)(x - 3) = x^2 - 3x + 3x - 3^2 = x^2 - 9$

Then multiply $(x^2 + 9)(x^2 - 9) = x^4 - 9x^2 + 9x^2 - 9^2 =$ $x^4 - 81$

Note: Factoring more complicated polynomials in the form $ax^2 + bx + c$ may require first factoring out a common monomial factor or require more than one factoring event.

- **Example:** Factor $2x^2 + 10x + 12$.

Factor out the common monomial factor, 2.

$$2x^2 + 10x + 12 = 2(x^2 + 5x + 6)$$

Factor the trinomial $(x^2 + 5x + 6)$.

Find sets of two values that, when multiplied, equal the first term of the trinomial.

Find sets of two values that, when multiplied, equal the last term of the trinomial.

Possible binomial sets are (there are no negative signs in the trinomial):

$$(x + 2)(x + 3)$$

$$(x + 6)(x + 1)$$

Choose the above sets such that the sum of the outer product and the inner product of the binomial is equal to the second term (the first degree term) of the trinomial. (Note that in this case the terms are positive.)

$(x + 2)(x + 3)$: outer product $= 3x$, inner product $= 2x$, sum $= 5x$

$(x + 6)(x + 1)$: outer product $= 1x$, inner product $= 6x$, sum $= 7x$

Because the second term of the trinomial is 5x, choose $(x + 2)(x + 3)$.

Multiplying the monomial factor 2 results in $(2)(x + 2)$ $(x + 3)$.

Multiply the resulting factors to check that the original trinomial is produced.

$$(2)(x + 2)(x + 3) = (2)(x^2 + 3x + 2x + 6)$$

$$= (2)(x^2 + 5x + 6) = 2x^2 + 10x + 12$$

Therefore, the factored form of $2x^2 + 10x + 12$ is

$$(2)(x + 2)(x + 3).$$

Chapter 6

Algebraic Fractions with Polynomial Expressions

6.1 Factoring and Reducing Algebraic Fractions

• When solving algebraic equations, it is generally beneficial to factor and reduce algebraic fractions when they are present.

• To factor and reduce an algebraic fraction, simply factor the numerator and the denominator separately, then reduce by comparing the numerator to the denominator and canceling common factors.

• **Example:** Factor and reduce $(2x^2 - 2)/(x + 1)$.

$$\frac{2x^2 - 2}{x + 1} = \frac{2(x^2 - 1)}{x + 1} = \frac{2(x + 1)(x - 1)}{x + 1} = \frac{2(x - 1)}{1}$$

$$= 2(x - 1)$$

Remember the difference of two squares:

$$(x^2 - 1) = (x + 1)(x - 1).$$

• **Example:** Factor and reduce $(x^2 + 5x + 6)/(x^2 + 4x + 4)$.

$$\frac{x^2 + 5x + 6}{x^2 + 4x + 4} = \frac{(x + 2)(x + 3)}{(x + 2)(x + 2)} = \frac{x + 3}{x + 2}$$

6.2 Multiplication of Algebraic Fractions

- The following method can be used for the multiplication of algebraic fractions.

- To multiply algebraic fractions, factor each numerator and denominator, reduce each fraction by canceling factors common to the numerator and denominator, then multiply the numerators with each other and the denominators with each other, and reduce the resulting fraction by canceling factors common to the numerator and denominator.

- **Example:** Multiply $(6y + 6z)/(4x^2 + 4x)$ and $(4x)/(2y + 2z)$.

$$\frac{6y+6z}{4x^2+4x} \times \frac{4x}{2y+2z} = \frac{(6)(y+z)}{(4x)(x+1)} \times \frac{4x}{2(y+z)}$$

$$= \frac{(3)(y+z)}{(2x)(x+1)} \times \frac{2x}{(y+z)} = \frac{(3)(y+z)(2x)}{(2x)(x+1)(y+z)} = \frac{3}{x+1}$$

- **Example:** Multiply $(x^2 + 5x + 6)/(6x)$ and $(24x^2)/(x^2 + 6x + 9)$.

$$\frac{x^2+5x+6}{6x} \times \frac{24x^2}{(x^2+6x+9)}$$

$$= \frac{(x+2)(x+3)}{6x} \times \frac{24x^2}{(x+3)(x+3)}$$

$$= \frac{(x+2)(x+3)(24x^2)}{6x(x+3)(x+3)} = \frac{(x+2)4x}{(x+3)} = \frac{4x^2+8x}{x+3}$$

6.3 Division of Algebraic Fractions

• The following method can be used for the division of algebraic fractions.

• To divide algebraic fractions, change the division format into a multiplication format by multiplying the first fraction by the reciprocal of the second fraction. Using the reciprocal of the second fraction provides a multiplication format. To multiply the first fraction by the reciprocal fraction, factor each numerator and denominator, reduce each fraction by canceling factors common to the numerator and denominator, then multiply the numerators with each other and the denominators with each other, and reduce the resulting fraction by canceling factors common to the numerator and denominator.

• **Example:** Divide:

$$(x^2 + 4x + 4)/(2x^2 - 8) \div (x^2 + 2x)/(4x - 8)$$

$$= \frac{x^2 + 4x + 4}{2x^2 - 8} \div \frac{(x^2 + 2x)}{4x - 8}$$

Set up as multiplication by taking reciprocal to second fraction.

$$\frac{x^2 + 4x + 4}{2x^2 - 8} \times \frac{4x - 8}{(x^2 + 2x)} = \frac{(x + 2)(x + 2)}{2(x^2 - 4)} \times \frac{4(x - 2)}{x(x + 2)}$$

$$= \frac{(x + 2)(x + 2)}{2(x + 2)(x - 2)} \times \frac{4(x - 2)}{x(x + 2)} = \frac{(x + 2)}{2(x - 2)} \times \frac{4(x - 2)}{x(x + 2)}$$

$$= \frac{(x + 2)(4)(x - 2)}{2(x - 2)(x)(x + 2)} = \frac{2}{x}$$

Remember from the difference of two squares:

$$(x^2 - 4) = (x + 2)(x - 2).$$

6.4 Addition and Subtraction of Algebraic Fractions

- The following methods can be used for the addition and subtraction of algebraic fractions, including fractions with common denominators and fractions with different denominators.

Fractions with Common Denominators

- To add or subtract fractions with common denominators, add or subtract the numerators, place the result over the common denominator, and reduce the resulting fraction by factoring and canceling factors common to the numerator and denominator.

- **Example:** Subtract $2x/(3(x - 2)) - 4/(3(x - 2))$.

$$\frac{2x}{3(x-2)} - \frac{4}{3(x-2)} = \frac{2x-4}{3(x-2)} = \frac{2(x-2)}{3(x-2)} = \frac{2}{3}$$

Fractions with Different Denominators

• To add or subtract fractions with different denominators, first form fractions that are equivalent to the original fractions but have a common denominator, then add or subtract the numerators. The procedure is as follows:

1. Find a common denominator. Determine the lowest common denominator that will contain each different factor to the highest power it occurs in any of the denominators. For example, if $(x + 1)$ occurs twice in one of the original denominators, it must occur twice in the common denominator. To find a common denominator, factor each denominator and then write multiples of each denominator until a common multiple is found. For example, if you have two denominators, 4 and 6, multiples of 4 are: 4, 8, 12, 16..., and multiples of 6 are: 6, 12, 18, 24, 30..., the lowest common multiple or lowest common denominator is 12.

2. Multiply each fraction by a different fraction (with its numerator equal to its denominator) to create new fractions with common denominators that are equivalent to the original fractions. (By having the numerator equal to the denominator, the value of each of the fractions remains unchanged (for example, $1/2 = 2/4$)). To determine what each multiplying fraction needs to be, compare the new common denominator with the denominators of each of the original fractions and create new fractions that contain the factors that are in the common denominator, but not in each original denominator.

3. After new equivalent fractions with common denominators have been created, add or subtract the numerators, place the result over the common denominator, and reduce the resulting fraction by factoring and canceling factors common to the numerator and denominator.

- **Example:** Add $(x^2 + 4x + 4)/(2x^2 - 8) + (4x - 8)/(x^2 + 2x)$.

Factor each fraction and reduce.

$$\frac{x^2 + 4x + 4}{2x^2 - 8} + \frac{4x - 8}{x^2 + 2x} = \frac{(x + 2)(x + 2)}{2(x + 2)(x - 2)} + \frac{4(x - 2)}{x(x + 2)}$$

$$= \frac{(x + 2)}{2(x - 2)} + \frac{4(x - 2)}{x(x + 2)}$$

Determine the lowest common denominator that will contain each different factor. (Remember, the lowest common denominator must contain all of the factors of each denominator as many times as they occur in a given denominator.) The factors of each denominator are $(2)(x - 2)$ and $(x)(x + 2)$. Therefore, the lowest common denominator is $(2)(x)(x + 2)(x - 2)$.

Find the multiplying fractions by comparing the new common denominator with the denominators of each of the original fractions, and create new fractions that contain the factors that are in the common denominator but not in each original fraction.

For the first fraction $\frac{(x+2)}{2(x-2)}$

The factors that are in the common denominator but not in the original denominator are (x)(x + 2). Therefore, the multiplying fraction for the first fraction is $\frac{x(x+2)}{x(x+2)}$

For the second fraction $\frac{4(x-2)}{x(x+2)}$

The factors that are in the common denominator but not in the original denominator are (2)(x – 2). Therefore, the multiplying fraction for the second fraction is $\frac{2(x-2)}{2(x-2)}$

Create new fractions with common denominators that are equivalent to the original fractions.

$$\frac{(x+2)(x)(x+2)}{(2)(x-2)(x)(x+2)}+\frac{(4)(x-2)(2)(x-2)}{(x)(x+2)(2)(x-2)}$$

After new equivalent fractions with common denominators have been created, add or subtract the numerators by combining like terms, place the result over the common denominator, and reduce the resulting fraction.

Add each numerator and place the result over the common denominator.

$$\frac{(x+2)(x)(x+2)+(4)(x-2)(2)(x-2)}{(2)(x-2)(x)(x+2)}$$

$$=\frac{x(x^2+4x+4)+8(x^2-4x+4)}{(2)(x-2)(x)(x+2)}$$

Combine like terms.

$$= \frac{x^3 + 4x^2 + 4x + 8x^2 - 32x + 32}{(2)(x-2)(x)(x+2)} = \frac{x^3 + 12x^2 - 28x + 32}{(2)(x-2)(x)(x+2)}$$

$$= \frac{x^3 + 12x^2 - 28x + 32}{(2x)(x^2 - 4)} = \frac{x^3 + 12x^2 - 28x + 32}{2x^3 - 8x}$$

Therefore, $(x^2 + 4x + 4)/(2x^2 - 8) + (4x - 8)/(x^2 + 2x)$ is

$$\frac{x^3 + 12x^2 - 28x + 32}{2x^3 - 8x}$$

Chapter 7

Solving Quadratic Polynomial Equations with One Unknown Variable

7.1 Defining and Solving Quadratic (Polynomial) Equations

• In this section, the quadratic equation is defined, and examples of quadratic equations and methods used to solve quadratic equations with one unknown variable are presented.

• *Quadratic equations* are non-linear equations with a second-degree term. If x is the unknown variable in a quadratic equation, its highest power is 2 and the variable would be expressed as x^2.

• Quadratic equations have the form $ax^2 + bx + c = 0$. In a quadratic equation the coefficient *a* can never be zero, but coefficient *b* or *c* can be zero. When quadratic equations are written, the coefficients for the second-degree term, the first-degree term, and the constant term are represented by a, b, and c, respectively. Examples of quadratic equations are:

$$ax^2 + bx + c = 0$$

$$3x^2 + 2x + 5 = 0$$

$$x^2 + 2 = 0$$

$$3x^2 + 2x = 0$$

• The following sections describe several methods that are used to solve quadratic equations with one unknown variable. In summary:

1. Factoring is useful for solving quadratic equations with two or three terms.

2. The quadratic formula is useful for solving any quadratic equation, particularly those with three terms.

3. The square root method is useful for solving quadratic equations with two terms if the b coefficient is zero, resulting in the absence of the first-degree term.

4. The method of completing the square is useful for solving quadratic equations, particularly those with a non-zero b coefficient. This is an alternate method to using the quadratic formula.

7.2 Using Factoring to Solve Quadratic Equations with One Unknown Variable

• The method of factoring is useful for solving many quadratic equations with one unknown variable that have the form of a binomial or a trinomial.

• To solve a quadratic equation using the method of factoring:

1. Express the equation in the form of a quadratic equation, $ax^2 + bx + c = 0$.

2. Factor the quadratic expression, $ax^2 + bx + c$.
3. Set each factor equal to zero and solve each resulting equation for the unknown variable.
4. Check the solutions by substituting into the original equation.

- **Example:** Solve $x^2 - 2x = 3$.

Express in the form, $ax^2 + bx + c = 0$.

$$x^2 - 2x - 3 = 0$$

Factor the quadratic expression, $ax^2 + bx + c$.

$$(x^2 - 2x - 3)$$

Find sets of two values that, when multiplied, equal the first term of the trinomial. Find sets of two values that when multiplied equal the last term of the trinomial.

Possible binomial sets are:

$$(x + 1)(x - 3)$$

$$(x - 1)(x + 3)$$

Choose the set where the sum of the outer product and the inner product of the binomial is equal to the second term (the first-degree term) of the trinomial.

$(x + 1)(x - 3)$: outer product $= -3x$, inner product $= 1x$, sum $= -2x$

$(x - 1)(x + 3)$: outer product $= 3x$, inner product $= -1x$, sum $= 2x$

The second term of the trinomial is $-2x$. Therefore, choose $(x + 1)(x - 3)$.

Check by multiplying the resulting binomials.

$$(x + 1)(x - 3) = x^2 - 3x + 1x - 3 = x^2 - 2x - 3$$

Therefore, the factors are $(x + 1)$ and $(x - 3)$.

Set each factor equal to zero.

$$(x + 1) = 0$$

$$(x - 3) = 0$$

Solve for the unknown variable in each equation.

$$x + 1 = 0$$

$$x = -1$$

$$x - 3 = 0$$

$$x = 3$$

Therefore, the solutions for $x^2 - 2x = 3$, are $x = -1$ and $x = 3$.

Check the solutions by substituting each one into the original equation, $x^2 - 2x = 3$

For $x = -1$,

$$(-1)^2 - 2(-1) = 3$$

$$1 + 2 = 3$$

$$3 = 3$$

For $x = 3$,

$$3^2 - 2(3) = 3$$

$$9 - 6 = 3$$

$$3 = 3$$

- **Example:** Solve $x^2 = 2x$.

Express the equation in the form of a quadratic equation, $ax^2 + bx + c = 0$.

$$x^2 - 2x = 0$$

Factor the quadratic expression, $ax^2 + bx + c$.

$$x(x - 2) = 0$$

The factors are x and $(x-2)$.

Set each factor equal to zero and solve each resulting equation for the unknown variable.

For the first factor:

$$x = 0$$

For the second factor:

$$x - 2 = 0$$

$$x = 2$$

Therefore, the solutions for $x^2 = 2x$, are $x = 0$ and $x = 2$.

Check the solutions by substituting into the original equation, $x^2 = 2x$

For $x = 0$,

$$0^2 = 2(0)$$

$$0 = 0$$

For $x = 2$,

$$2^2 = 2(2)$$

$$4 = 4$$

7.3 Using the Quadratic Formula to Solve Quadratic Equations with One Unknown Variable

- The *quadratic formula* can be used to find the solution to any quadratic equation, particularly equations that have the form of a trinomial. The quadratic formula is worth memorizing.

The quadratic formula is:

$$x = \frac{-b \pm \sqrt{b^2 - 4ac}}{2a}$$

- The expression inside the square root is called the *discriminant*, and can be used to determine if the equation will have real or imaginary roots.

If $b^2 - 4ac > 0$, the equation has 2 distinct real roots.

If $b^2 - 4ac < 0$, the equation has 2 imaginary roots.

If $b^2 - 4ac = 0$, the equation has 1 real root.

- The quadratic formula can be derived using the method of completing the square.

- To solve a quadratic equation using the quadratic formula:

1. Express the equation in the form of a quadratic equation, $ax^2 + bx + c = 0$.
2. Identify the values for the coefficients a, b, and c.
3. Substitute the values for a, b, and c into the quadratic formula.
4. Reduce the resulting equation by performing the indicated arithmetic operations and simplify the radical.
5. Check the solutions by substituting into the original equation.

- **Example:** Solve $3x^2 + 2 = -5x$.

Express the equation in the form of a quadratic equation, $ax^2 + bx + c = 0$.

$$3x^2 + 5x + 2 = 0$$

Identify the values for the coefficients a, b, and c.

$$a = 3, b = 5, c = 2$$

Substitute the values for a, b, and c into the quadratic formula.

$$x = \frac{-5 \pm \sqrt{5^2 - 4(3)(2)}}{2(3)}$$

Reduce the resulting equation by performing the indicated arithmetic operations and simplify the radical.

$$x = \frac{-5 \pm \sqrt{25 - 24}}{6}$$

$$x = \frac{-5 \pm \sqrt{1}}{6}$$

Because of the ± sign, both + and – must be accounted for.

Also, $\sqrt{1} = 1$.

$$x = \frac{-5 + 1}{6} \text{ and } x = \frac{-5 - 1}{6}$$

Reduce each equation.

$$x = \frac{-4}{6} = \frac{-2}{3} \text{ and } x = \frac{-6}{6} = -1$$

Therefore, the solutions for $3x^2 + 2 = -5x$ are $x = -2/3$ and $x = -1$

Check the solutions by substituting into the original equation, $3x^2 + 2 = -5x$.

For $x = -2/3$,

$3(-2/3)^2 + 2 = -5(-2/3)$

$3(4/9) + 2 = 10/3$

$12/9 + 2 = 10/3$

$4/3 + 2 = 10/3$

Find the common denominator, because $2/1 = 6/3$,

$4/3 + 6/3 = 10/3$

$10/3 = 10/3$

For $x = -1$,

$3(-1)^2 + 2 = -5(-1)$

$3(1) + 2 = 5$

$5 = 5$

7.4 Using the Square Root Method to Solve Quadratic Equations with One Unknown Variable

- The *square root method* is useful for solving quadratic equations with two terms if the b coefficient is zero, resulting in the absence of the first-degree term. When the b coefficient is zero, the form of the quadratic equation changes from $ax^2 + bx + c = 0$ to $ax^2 + c = 0$.

- To solve a quadratic equation using the square root method:

1. Isolate the x^2 variable on one side of the equal sign (not the whole term).
2. Take the square root of both sides of the equation. This transforms x^2 into x, because $\sqrt{x^2} = \pm x$.
3. Simplify the radicals.
4. Check the solutions by substituting into the original equation.

- **Example:** Solve $3x^2 + 4 = 31$.

Subtract 4 from both sides of the equation.

$$3x^2 = 27$$

To isolate the x^2 variable on one side of the equal sign, divide both sides by 3.

$$x^2 = 27/3$$

$$x^2 = 9$$

Take the square root of both sides of the equation.

$$\sqrt{x^2} = \sqrt{9}$$

$$x = \pm 3$$

Therefore, the solutions to $3x^2 + 4 = 31$ are $x = 3$ and $x = -3$.

Check the solutions by substituting into the original equation $3x^2 + 4 = 31$.

For $x = 3$,

$$3(3)^2 + 4 = 31$$

$$3(9) + 4 = 31$$

$$27 + 4 = 31$$

$$31 = 31$$

For $x = -3$,

$$3(-3)^2 + 4 = 31$$

$$3(9) + 4 = 31$$

$$27 + 4 = 31$$

$$31 = 31$$

7.5 Using the Method of Completing the Square to Solve Quadratic Equations with One Unknown Variable

• The *method of completing the square* is an extension of the square root method that is useful for solving quadratic equations, particularly equations with a non-zero b coefficient. It is an alternate method to the quadratic formula described previously.

• To solve a quadratic equation using the method of completing the square:

1. Express the equation in the form $x^2 + bx = c$. To obtain this form it may be necessary to divide each term by the coefficient a, $ax^2/a + bx/a = c/a$.

2. Complete the square by finding one-half of the coefficient b, $(b/2)$. Square one-half of coefficient b, $(b/2)^2$, and add the result to each side of the equation.

3. Factor the resulting perfect square trinomial expression into a binomial-squared, and combine like terms.

4. Solve using the square root method, by isolating the x^2 variable on one side of the equal sign, taking the square root of both sides of the equation, and simplifying the radicals.

5. Check the solutions by substituting into the original equation.

- **Example:** Solve $x^2 - 2x - 10 = -7$.

Express the equation in the form, $x^2 + bx = c$.

Add 10 to both sides.

$$x^2 - 2x = 3$$

Complete the square by finding one-half of the coefficient b, ($b=-2$). Square one-half of the coefficient b, $(b/2)^2$, and add the result to each side of the equation.

$$(b/2)^2 = (-2/2)^2 = (-1)^2 = 1$$

Therefore:

$$x^2 - 2x + 1 = 3 + 1$$

$$x^2 - 2x + 1 = 4$$

Factor the resulting perfect square trinomial expression into a binomial-squared, and combine like terms.

$$x^2 - 2x + 1 = 4$$

$$(x - 1)(x - 1) = 4$$

Check factoring:

$$(x - 1)(x - 1) = x^2 - 1x - 1x + 1 = x^2 - 2x + 1$$

Insert back into the equation.

$$(x - 1)(x - 1) = 4$$

Equivalently:

$$(x - 1)^2 = 4$$

Solve using the square root method.

Isolate the x^2 variable on one side of the equal sign.

$$(x - 1)^2 = 4$$

The x^2 variable is represented by $(x - 1)^2$.

Take the square root of both sides of the equation and simplify the radicals.

$$\sqrt{(x - 1)^2} = \sqrt{4}$$

$$x - 1 = \pm 2$$

Remember, $\sqrt{x^2} = \pm x$.

Solve for x for +2 and for −2.

For +2,

$$x - 1 = 2$$

$$x = 3$$

For −2,

$$x - 1 = -2$$

$$x = -1$$

Therefore, the solutions to $x^2 - 2x - 10 = -7$ are $x = 3$ and $x = -1$.

Check the solutions by substituting into the original equation, $x^2 - 2x - 10 = -7$.

For $x = 3$,

$$(3)^2 - 2(3) - 10 = -7$$

Multiply, and add 7 to both sides.

$$9 - 6 - 3 = 0$$

Remember to subtract in order.

$$0 = 0$$

For $x = -1$,

$$(-1)^2 - 2(-1) - 10 = -7$$

Multiply, and add 7 to both sides.

$$1 + 2 - 3 = 0$$

$$0 = 0$$

Chapter 8

Solving Systems of Linear Equations with Two or Three Unknown Variables

8.1 Solving Systems of Linear Equations with Two or More Unknown Variables

• There are several methods that are used to solve systems of linear equations for two or more unknown variables. The number of unknown variables that must be determined dictates the number of equations that must be solved.

• To find the solutions for two unknown variables, there must be two equations that contain the two unknown variables. To find solutions for three unknown variables, there must be three equations that contain the three unknown variables. To find solutions for n unknown variables, there must be n equations that contain the unknown variables.

• To solve word problems with unknown variables, translate the words in the problem into algebraic equations containing unknown variables and determine which method described in this chapter is most appropriate to solve the set of algebraic equations.

Setting Up Word Problems Using Two Variables

• **Example:** What is the value of two numbers that, when added, the result is 20, and when subtracted, the result is 10?

Let x represent one number and y represent the other number.

If the sum is 20, then $x + y = 20$.

If the difference is 10, then $x - y = 10$.

These equations are solved using the elimination method discussed in the next section.

- **Example:** Suppose you are a chemist and need 50 liters of a 10% acid solution, but you only have 8% and 14% acid solutions in your laboratory. How many liters of 8% and 14% acid solutions must be mixed together to create 50 liters of a 10% solution of acid?

Let x represent the number of liters needed of the 8% acid solution. Let y represent the number of liters needed of the 14% acid solution.

For 50 liters, x liters + y liters = 50 liters

For a 10% solution, (8%)(x liters) + (14%)(y liters) = (10%)(50 liters).

These two equations are solved using the method of determinants later in this chapter.

- The following sections describe several methods that are used to solve systems of two equations with two unknowns and three equations with three unknowns.

Methods for Two Equations and Two Unknowns

Elimination Method

Substitution Method

Method of Determinants

Methods for Three Equations and Three Unknowns

Elimination Method

Substitution Method

Matrix Method

Method of Determinants of a Square Matrix

8.2 Using the Elimination Method to Solve Systems of Linear Equations with Two Unknown Variables

• The *elimination method* is designed to solve for the unknown variables by eliminating the first variable by adding the two equations together and then solving the resulting equation for the second unknown variable. The first unknown variable can then be solved for by substituting the second (now known) variable into one of the original equations.

• To solve two equations using the elimination method:

1. Evaluate whether adding the two equations together will eliminate either of the terms containing an unknown variable.

2. If adding the two equations does not eliminate one of the terms, multiply each term in one equation by a number that will allow one of the terms to be eliminated when the two equations are added.

3. Add the equations to eliminate one of the terms containing an unknown variable, and solve the resulting equation for the remaining unknown variable.

4. Substitute the value for the known variable found in the resulting equation in the preceding step into either one of the original equations, and solve for the other unknown variable.

5. Check the solutions by substituting them into each of the original equations.

- **Example:** Solve the equations $x - y = 4$ and $3x + 2y = 7$.

Evaluate whether adding the two equations together will eliminate the x term or y term in either equation.

Adding the x terms: $x + 3x = 4x$

No, does not eliminate.

Adding the y terms: $-y + 2y = y$

No, does not eliminate.

If adding the two equations does not eliminate the x or y terms, multiply each term in one of the equations by a number that will allow one of the terms to be eliminated once the two equations are added. Multiply $x - y = 4$ by 2 to eliminate the y terms when the equations are added.

$$2(x - y = 4) = 2x - 2y = 8$$

Add the equations to eliminate one of the terms.

$$
\begin{array}{r}
2x - 2y = 8 \\
+\ 3x + 2y = 7 \\
\hline
5x + 0 = 15
\end{array}
$$

Solve the resulting equation for the remaining unknown variable.

$5x = 15$

$x = 15/5$

$x = 3$

Substitute $x = 3$ into either one of the original equations and solve for y.

$x - y = 4$

$3 - y = 4$

$-y = 4 - 3$

$-y = 1$

$y = -1$

Therefore, the solutions to $x - y = 4$ and $3x + 2y = 7$ are $x = 3$ and $y = -1$.

Check the solutions by substituting them into each of the original equations $x - y = 4$ and $3x + 2y = 7$.

The first equation, $x - y = 4$

$$3 - (-1) = 4$$

$$3 + 1 = 4$$

$$4 = 4$$

The second equation, $3x + 2y = 7$

$$3(3) + 2(-1) = 7$$

$$9 - 2 = 7$$

$$7 = 7$$

- **Example:** Remember the example in the previous section in which there are two numbers whose sum is 20 and whose difference is 10. What are these numbers?

Let x represent one number.

Let y represent the other number.

If the sum is 20, then $x + y = 20$.

If the difference is 10, then $x - y = 10$.

Solve using the elimination method.

Add the equations to eliminate one variable and solve for the remaining variable.

$$\begin{array}{rcl} x + y & = & 20 \\ + \; x - y & = & 10 \\ \hline 2x & = & 30 \end{array}$$

$2x = 30$

$x = 30/2$

$x = 15$

Substitute $x = 15$ into one of the original equations and solve for y.

$x - y = 10$

$15 - y = 10$

$15 - 10 = y$

$y = 5$

Therefore, the solutions to $x + y = 20$ and $x - y = 10$ are $x = 15$ and $y = 5$.

Check by substituting into the original equations.

The first equation, $x + y = 20$

$15 + 5 = 20$

$20 = 20$

The second equation, $x - y = 10$

$15 - 5 = 10$

$10 = 10$

8.3 Using the Substitution Method to Solve Systems of Linear Equations with Two Unknown Variables

- The *substitution method* is designed to solve for the two unknown variables by isolating one of the variables in one equation and substituting the expression the variable is equal to into the other equation, solving the resulting equation for the other unknown variable, then substituting the known variable into one of the original equations to solve for the first unknown variable.

- To solve two equations using the substitution method:

1. Solve either one of the equations for one of the unknown variables by isolating that variable on one side of the equal sign (the other unknown variable will be present on the other side of the equal sign).

2. Substitute the expression that is equal to the isolated unknown variable into the other equation.

3. Solve the resulting equation for the unknown variable that is present.

4. Substitute the value for the variable found in the previous step into either one of the original equations and solve for the still unknown variable.

5. Check the solutions by substituting them into each of the original equations.

- **Example:** Solve $2x - y = 4$ and $2x + 4y = 4$.

Solve either one of the equations for one of the unknown variables by isolating that variable on one side of the equal sign.

$$2x - y = 4$$

$$2x - 4 = y$$

Substitute the expression for the isolated y variable into the other equation.

$$2x + 4y = 4$$

$$2x + 4(2x - 4) = 4$$

Solve the resulting equation for the x variable. (Multiply 4 with the binomial, then add 16 to both sides.)

$$2x + 4(2x - 4) = 4$$

$$2x + 8x - 16 = 4$$

$$10x = 20$$

Divide both sides by 10.

$$x = 20/10$$

$$x = 2$$

Substitute $x = 2$ into either one of the original equations and solve for y.

$$2x - y = 4$$

$$2(2) - y = 4$$

$$4 - y = 4$$

Add y to both sides and subtract 4 from both sides.

$$4 - 4 = y$$

$$0 = y$$

Therefore, the solutions to $2x - y = 4$ and $2x + 4y = 4$ are $x = 2$ and $y = 0$.

Check the solutions by substituting them into each of the original equations.

The first equation, $2x - y = 4$

$$2(2) - 0 = 4$$

$$4 = 4$$

The second equation, $2x + 4y = 4$

$$2(2) + 4(0) = 4$$

$$4 + 0 = 4$$

$$4 = 4$$

8.4 Using the Method of Determinants to Solve Systems of Two Linear Equations with Two Unknown Variables

- The *method of determinants* is used to solve systems of two linear equations with two unknown variables by writing the coefficients of the equations in a matrix format, calculating the determinant of each matrix, and then solving for the unknown variables using the determinants.

- A rectangular array of numbers configured in rows and columns is called a *matrix*. For example:

$$\begin{vmatrix} 1 & 2 \\ 3 & 4 \end{vmatrix}$$

- The matrix format provides a tool that can be used to find the value of unknown variables in systems of equations.

- In the 2 by 2 matrix:

$$\begin{vmatrix} a_1 & b_1 \\ a_2 & b_2 \end{vmatrix}$$

The *determinant* of a 2 by 2 matrix is $(a_1)(b_2) - (a_2)(b_1)$.

- For example, find the determinant.

$$D = \begin{vmatrix} 1 & 2 \\ 3 & 4 \end{vmatrix} = (1)(4) - (3)(2) = 4 - 6 = -2$$

- To use the *method of determinants* to solve two equations with two unknown variables, *Cramer's rule* is applied. To solve two equations for x and y:

1. Write the two equations in the form:

$$a_1 x + b_1 y = c_1$$

$$a_2 x + b_2 y = c_2$$

x and y represent unknown variables and a, b, and c represent known coefficients.

2. Create three matrices of coefficients (D, D_x, and D_y), and calculate the determinants of each matrix as follows:

$$D = \begin{vmatrix} a_1 & b_1 \\ a_2 & b_2 \end{vmatrix} = (a_1)(b_2) - (a_2)(b_1)$$

$$D_x = \begin{vmatrix} c_1 & b_1 \\ c_2 & b_2 \end{vmatrix} = (c_1)(b_2) - (c_2)(b_1)$$

$$D_y = \begin{vmatrix} a_1 & c_1 \\ a_2 & c_2 \end{vmatrix} = (a_1)(c_2) - (a_2)(c_1)$$

Each matrix is made up of the coefficients in the equations. Also, D cannot be zero.

3. Solve for x and y, using the following definitions:

$$x = D_x/D$$

$$y = D_y/D$$

4. Check the results, substitute the values for x and y into the original two equations.

• Example (from the first section of this chapter): Suppose you are a chemist and need 50 liters of a 10% acid solution, but you only have 8% and 14% acid solutions in your laboratory. How many liters of 8% and 14% acid solutions must be mixed together to create 50 liters of a 10% solution of acid?

Let x represent the number of liters needed of the 8% acid solution.

Let y represent the number of liters needed of the 14% acid solution.

For 50 liters, x liters + y liters = 50 liters

For a 10% solution, (8%)(x liters) + (14%)(y liters) = (10%)(50 liters)

Write the equations in the form $a_1x + b_1y = c_1$ and $a_2x + b_2y = c_2$.

$$x + y = 50$$

$$(0.08)(x) + (0.14)(y) = (0.1)(50)$$

Multiply the second equation by 100 to eliminate the decimal.

$$(8)(x) + (14)(y) = (10)(50)$$

The equations become:

$$x + y = 50$$

$$8x + 14y = 500$$

The coefficients a_1, b_1, and c_1 in the first equation are 1, 1, and 50, respectively.

The coefficients a_2, b_2, and c_2 in the second equation are 8, 14, and 500 respectively.

Using Cramer's rule:

$$x = D_x/D$$

$$y = D_y/D$$

Find the determinants.

$$D = \begin{vmatrix} a_1 & b_1 \\ a_2 & b_2 \end{vmatrix} = (a_1)(b_2) - (a_2)(b_1)$$

$$D = \begin{vmatrix} 1 & 1 \\ 8 & 14 \end{vmatrix} = (1)(14) - (8)(1) = 14 - 8 = 6$$

$$D_x = \begin{vmatrix} c_1 & b_1 \\ c_2 & b_2 \end{vmatrix} = (c_1)(b_2) - (c_2)(b_1)$$

$$D_x = \begin{vmatrix} 50 & 1 \\ 500 & 14 \end{vmatrix} = (50)(14) - (500)(1) = 700 - 500 = 200$$

$$D_y = \begin{vmatrix} a_1 & c_1 \\ a_2 & c_2 \end{vmatrix} = (a_1)(c_2) - (a_2)(c_1)$$

$$D_y = \begin{vmatrix} 1 & 50 \\ 8 & 500 \end{vmatrix} = (1)(500) - (8)(50) = 500 - 400 = 100$$

Therefore, x and y are calculated as:

$$x = D_x/D = 200/6 = 100/3$$

$$y = D_y/D = 100/6 = 50/3$$

Therefore, the solutions to $x + y = 50$ and $(0.08)(x) + (0.14)(y) = (0.1)(50)$ are $x = 100/3$ (or approximately 33.33) liters of the 8% acid solution and $y = 50/3$ (or approximately 16.67) liters of the 14% acid solution.

Check by substituting into the original equations.

The first equation:

$$x + y = 50$$

$$100/3 + 50/3 = 50$$

$$150/3 = 50$$

$$50 = 50$$

The second equation:

$$(0.08)(x) + (0.14)(y) = (0.1)(50)$$

$$(0.08)(100/3) + (0.14)(50/3) = (0.1)(50)$$

$$8/3 + 7/3 = 5$$

$$15/3 = 5$$

$$5 = 5$$

8.5 Solving Systems of Three Linear Equations with Three Unknown Variables

• There are several methods that are used to solve systems of three linear equations with three unknown variables.

• Because the number of unknown variables that must be determined dictates the number of equations that must be solved, if there are three unknown variables in a problem, there must be three equations containing the three unknown variables that describe the problem.

• To solve for three unknown variables, an expanded version of the following methods can be used. These methods are described in detail in the following sections.

Elimination Method

Substitution Method

Matrix Method

Method of Determinants of a Square Matrix

8.6 Using the Elimination Method to Solve Systems of Three Linear Equations with Three Unknown Variables

- The *elimination method* used to solve three equations for three unknown variables is an expanded version of the elimination method used to solve two equations for two unknown variables described earlier in this chapter.

- To solve three equations using the elimination method:

1. Evaluate whether adding two of the equations together will eliminate one of the variables.

2. If adding two equations does not eliminate one of the variables, multiply each term in one equation by a number that will allow one of the terms to be eliminated if two equations are added.

3. Add the equations to eliminate one of the variables.

4. Solve the resulting equation for the remaining unknown variable, and substitute the value for that variable into one of the other equations and solve for a second unknown variable.

5. Substitute the value for the second variable found in the resulting equation into one of the equations and solve for the third unknown variable.

6. Check the solutions by substituting them into each of the original equations.

- **Example:** Solve the following equations for x, y, and z.

$$2x + 3y + 3z = 4$$

$$-2x + 3y - 4z = 5$$

$$4x + 2y + 5z = 4$$

Add the first two equations to eliminate x from the second equation.

$$\begin{array}{r} 2x + 3y + 3z = 4 \\ + \; -2x + 3y - 4z = 5 \\ \hline 6y - 1z = 9 \end{array}$$

Add (−2) times the first equation to the third equation to eliminate x from the third equation.

$$\begin{array}{r} -4x - 6y - 6z = -8 \\ + \quad 4x + 2y + 5z = 4 \\ \hline -4y - 1z = -4 \end{array}$$

Resulting in the three equations:

$$2x + 3y + 3z = 4$$

$$6y - 1z = 9$$

$$-4y - 1z = -4$$

Multiply the third equation by -1 and add it to the second equation to eliminate z.

$$\begin{array}{rrcrcr} & 6y & - & 1z & = & 9 \\ + & 4y & + & 1z & = & 4 \\ \hline & 10y & & & = & 13 \end{array}$$

Solve $10y = 13$ for y. Divide both sides by 10.

$y = 13/10$

$y = 1.3$

Solve for z by substituting $y = 1.3$ into the second (reduced) equation.

$6y - z = 9$

$6(1.3) - z = 9$

$7.8 - z = 9$

Subtract 9 from both sides, and add z to both sides.

$z = -1.2$

Solve for x by substituting $y = 1.3$ and $z = -1.2$ into the first equation.

$2x + 3y + 3z = 4$

$2x + 3(1.3) + 3(-1.2) = 4$

$2x + 3.9 - 3.6 = 4$

$2x + 0.3 = 4$

Subtract 0.3 from both sides.

$$2x = 3.7$$

Divide both sides by 2.

$$x = 1.85$$

Therefore, the solutions to the equations:

$$2x + 3y + 3z = 4$$

$$-2x + 3y - 4z = 5$$

$$4x + 2y + 5z = 4$$

are $x = 1.85$, $y = 1.3$, and $z = -1.2$.

Check solutions by substituting the variables into each original equation.

The first equation:

$$2x + 3y + 3z = 4$$

$$2(1.85) + 3(1.3) + 3(-1.2) = 4$$

$$3.7 + 3.9 - 3.6 = 4$$

$$7.6 - 3.6 = 4$$

$$4 = 4$$

The second equation:

$$-2x + 3y - 4z = 5$$

$$-2(1.85) + 3(1.3) - 4(-1.2) = 5$$

$$-3.7 + 3.9 + 4.8 = 5$$

$$0.2 + 4.8 = 5$$

$$5 = 5$$

The third equation:

$$4x + 2y + 5z = 4$$

$$4(1.85) + 2(1.3) + 5(-1.2) = 4$$

$$7.4 + 2.6 - 6 = 4$$

$$10 - 6 = 4$$

$$4 = 4$$

8.7 Using the Substitution Method to Solve Systems of Three Linear Equations with Three Unknown Variables

• The *substitution method* used to solve three equations for three unknown variables is an expanded version of the substitution method used to solve two equations for two unknown variables described earlier in this chapter.

• To solve three equations using the substitution method, isolate selected variables and substitute them into the equations. The procedure is:

1. Isolate one unknown variable in one of the equations on one side of the equal sign (the other unknown variables will be present on the other side of the equal sign), substitute the expression into another equation, and solve the resulting equation for a second unknown variable.

2. Continue to substitute expressions or values for the variables into the original and reduced equations and solve for the unknown variables until all three variables are known.

3. Check the solutions by substituting them into each of the original equations.

• **Example:** Solve the following three equations using the substitution method.

$$4x + 2y - 2z = -2$$

$$3x + 2y + z = -1$$

$$2x - 6y + 6z = 6$$

Solve the first equation for y. Subtract 4x and add 2z to both sides.

$$4x + 2y - 2z = -2$$

$$2y = 2z - 4x - 2$$

Divide both sides by 2.

$$y = (1/2)(2z - 4x - 2)$$

$$y = z - 2x - 1$$

Substitute the expression for y into the second equation and reduce.

$$3x + 2(z - 2x - 1) + z = -1$$

$$3x + 2z - 4x - 2 + z = -1$$

Combine like terms.

$$-x + 3z - 2 = -1$$

Add 2 to both sides.

$$-x + 3z = 1$$

Substitute the expression for y into the third equation and solve for x.

$$2x - 6(z - 2x - 1) + 6z = 6$$

$$2x - 6z + 12x + 6 + 6z = 6$$

Combine like terms.

$$14x + 6 = 6$$

Subtract 6 from both sides.

$$14x = 0$$

Divide both sides by 14.

$$x = 0$$

To solve for z, substitute $x = 0$ into the reduced equation, $-x + 3z = 1$.

$$-x + 3z = 1$$

$$-0 + 3z = 1$$

$$3z = 1$$

Divide both sides by 3.

$$z = 1/3$$

Substitute $x = 0$ and $z = 1/3$ into the first equation (with y isolated) and solve for y.

$$y = z - 2x - 1$$

$$y = 1/3 - 2(0) - 1$$

$$y = 1/3 - 1$$

$$y = 1/3 - 3/3$$

$$y = -2/3$$

Therefore, the solutions to the three equations,

$$4x + 2y - 2z = -2$$

$$3x + 2y + z = -1$$

$$2x - 6y + 6z = 6$$

are $x = 0$, $y = -2/3$, and $z = 1/3$.

Check solutions by substituting them into each equation.

The first equation:

$$4x + 2y - 2z = -2$$

$$4(0) + 2(-2/3) - 2(1/3) = -2$$

$$-4/3 - 2/3 = -2$$

$$-6/3 = -2$$

$$-2 = -2$$

The second equation:

$$3x + 2y + z = -1$$

$$3(0) + 2(-2/3) + 1/3 = -1$$

$$-4/3 + 1/3 = -1$$

$$-3/3 = -1$$

$$-1 = -1$$

The third equation:

$$2x - 6y + 6z = 6$$

$$2(0) - 6(-2/3) + 6(1/3) = 6$$

$$12/3 + 6/3 = 6$$

$$4 + 2 = 6$$

$$6 = 6$$

(Note: Different substitutions in a different order can be made.)

8.8 Using the Matrix Method to Solve Systems of Three Linear Equations with Three Unknown Variables

• The *matrix method* is designed to solve three equations with three unknown variables by: transforming the equations into a matrix; performing operations on the matrix until an "upper triangular matrix" is formed; transforming the matrix back into three equations; and solving the resulting equations for the unknown variables using substitution.

Expanding on the above procedure:

1. Transform the equations into a matrix by writing the coefficients of the equations into a matrix format as depicted below with a on the left, b next, etc.

2. To create an upper triangular matrix, add multiples of the coefficient rows to each other until an upper triangular matrix is formed. This step in the process involves:
   a) multiplying a row through by a non−zero constant;
   b) interchanging the position of two rows (so that a zero coefficient will result in the lower left corner of the matrix); and c) adding the multiple of the row to another row which results in a zero coefficient in the lower left of the upper triangular matrix. See below.

3. Once the upper triangular matrix is formed, transform the coefficient matrix back into the form of the three equations by writing the three equations with their resulting coefficients as depicted below.

4. Solve the resulting equations for the unknown variables using substitution.

5. Check by substituting the unknown variables into the original equations.

- To use the matrix method, write the equations in the form:

$$a_1x + b_1y + c_1z = d_1$$

$$a_2x + b_2y + c_2z = d_2$$

$$a_3x + b_3y + c_3z = d_3$$

The corresponding matrix of coefficients is:

$$\begin{vmatrix} a_1 & b_1 & c_1 & d_1 \\ a_2 & b_2 & c_2 & d_2 \\ a_3 & b_3 & c_3 & d_3 \end{vmatrix}$$

To create the *upper triangular matrix*, the a_2, a_3, and b_3 coefficients must be converted to zeros and the a_1, b_2, and c_3 coefficients must be converted to ones.

$$\begin{vmatrix} 1 & b_1 & c_1 & d_1 \\ 0 & 1 & c_2 & d_2 \\ 0 & 0 & 1 & d_3 \end{vmatrix}$$

The known variable (from the third equation in the reduced matrix) can be substituted back into a reduced equation to solve for the next variable and then both known variables can be substituted into an original equation to find the third variable.

Check the results by substituting the values into the original equations.

- **Example:** Solve the following three equations, which are written in the proper form.

$$x + y + z = -1$$

$$2x - y + z = 0$$

$$-x + y - z = -2$$

The corresponding matrix of coefficients is:

$$\begin{vmatrix} 1 & 1 & 1 & -1 \\ 2 & -1 & 1 & 0 \\ -1 & 1 & -1 & -2 \end{vmatrix}$$

To reduce to an upper triangular matrix, a_1 must be 1 (which it is already).

Next, zeros must occupy the a_2 and a_3 positions. To achieve this, multiply row 1 by -2 and add it to row 2 to make the a_2 position zero.

-2	-2	-2	2	$-2 \times$ row 1
$+\ 2$	-1	1	0	row 2
0	-3	-1	2	new row 2

Add row 1 and row 3 to make the a_3 position zero.

	1	1	1	−1	row 1
+	−1	1	−1	−2	row 3
	0	2	0	−3	new row 3

The new matrix becomes:

$$\begin{vmatrix} 1 & 1 & 1 & -1 \\ 0 & -3 & -1 & 2 \\ 0 & 2 & 0 & -3 \end{vmatrix}$$

Switch the second and third rows.

$$\begin{vmatrix} 1 & 1 & 1 & -1 \\ 0 & 2 & 0 & -3 \\ 0 & -3 & -1 & 2 \end{vmatrix}$$

Add row 2 and row 3.

	0	2	0	−3	row 2
+	0	−3	−1	2	row 3
	0	−1	−1	−1	new row 2

Multiply (−1) times new row 2 to eliminate the negative sign so that the b_2 coefficient will be 1.

$$\begin{vmatrix} 1 & 1 & 1 & -1 \\ 0 & 1 & 1 & 1 \\ 0 & -3 & -1 & 2 \end{vmatrix}$$

To make the b_3 coefficient to be zero, add 3 times row 2 to row 3.

$$\begin{array}{cccccl} & 0 & 3 & 3 & 3 & 3 \times \text{row } 2 \\ + & 0 & -3 & -1 & 2 & \text{row } 3 \\ \hline & 0 & 0 & 2 & 5 & \text{new row } 3 \end{array}$$

The new matrix becomes:

$$\begin{vmatrix} 1 & 1 & 1 & -1 \\ 0 & 1 & 1 & 1 \\ 0 & 0 & 2 & 5 \end{vmatrix}$$

To make the c_3 coefficient to be 1, divide row 3 by 2.

$$\begin{vmatrix} 1 & 1 & 1 & -1 \\ 0 & 1 & 1 & 1 \\ 0 & 0 & 1 & 5/2 \end{vmatrix}$$

This is the upper triangular matrix corresponding to the three equations. The three equations taken from the upper triangular matrix are:

$$1x + 1y + 1z = -1$$

$$1y + 1z = 1$$

$$1z = 5/2$$

Therefore, $z = 5/2$.

Substitute $z = 5/2$ into the second reduced equation.

$$1y + 1z = 1$$

$$y + 5/2 = 1$$

Subtract 5/2 from both sides.

$$y = 1 - (5/2)$$

$$y = 2/2 - 5/2$$

$$y = -3/2$$

Substitute $y = -3/2$ and $z = 5/2$ into the first equation.

$$x + y + z = -1$$

$$x + (-3/2) + (5/2) = -1$$

$$x + 2/2 = -1$$

$$x + 1 = -1$$

Subtract 1 from both sides.

$$x = -1 - 1$$

$$x = -2$$

Therefore, the solutions to:

$$x + y + z = -1$$

$$2x - y + z = 0$$

$$-x + y - z = -2$$

are $x = -2$, $y = -3/2$, and $z = 5/2$.

Check by substituting the solutions into the original three equations.

The first equation:

$$x + y + z = -1$$

$$-2 + -3/2 + 5/2 = -1$$

$$-2 + 2/2 = -1$$

$$-2 + 1 = -1$$

$$-1 = -1$$

The second equation:

$$2x - y + z = 0$$

$$2(-2) - (-3/2) + 5/2 = 0$$

$$-4 + 8/2 = 0$$

$$-4 + 4 = 0$$

$$0 = 0$$

The third equation:

$$-x + y - z = -2$$

$$-(-2) + (-3/2) - 5/2 = -2$$

$$2 - 8/2 = -2$$

$$2 - 4 = -2$$

$$-2 = -2$$

8.9 Using the Method of Determinants of a Square Matrix to Solve Systems of Three Linear Equations with Three Unknown Variables

- A *square matrix* has an equal number of rows and columns resulting in a square shape. For example:

$$\begin{vmatrix} a_1 & b_1 & c_1 \\ a_2 & b_2 & c_2 \\ a_3 & b_3 & c_3 \end{vmatrix}$$

- The determinant of a square, 3 by 3 matrix is:

$$\begin{vmatrix} a_1 & b_1 & c_1 \\ a_2 & b_2 & c_2 \\ a_3 & b_3 & c_3 \end{vmatrix}$$

$$= a_1 \times \begin{vmatrix} b_2 & c_2 \\ b_3 & c_3 \end{vmatrix} - a_2 \times \begin{vmatrix} b_1 & c_1 \\ b_3 & c_3 \end{vmatrix} + a_3 \times \begin{vmatrix} b_1 & c_1 \\ b_2 & c_2 \end{vmatrix}$$

$$= a_1(b_2c_3 - b_3c_2) - a_2(b_1c_3 - b_3c_1) + a_3(b_1c_2 - b_2c_1)$$

Remember from the definition of a determinant of a 2 by 2 matrix:

$$\begin{vmatrix} a_1 & b_1 \\ a_2 & b_2 \end{vmatrix} = (a_1)(b_2) - (a_2)(b_1)$$

- To solve three first–degree equations with three unknown variables using the *method of determinants of a square matrix, Cramer's rule* can be applied.

- The following procedure is used to solve for three variables:

1. Write the three equations in the form:

$$a_1x + b_1y + c_1z = d_1$$

$$a_2x + b_2y + c_2z = d_2$$

$$a_3x + b_3y + c_3z = d_3$$

2. Create four square 3 by 3 matrices of coefficients (D, D_x, D_y, and D_z), and calculate the determinants of each matrix as follows:

$$D = \begin{vmatrix} a_1 & b_1 & c_1 \\ a_2 & b_2 & c_2 \\ a_3 & b_3 & c_3 \end{vmatrix}$$

$$= a_1 \times \begin{vmatrix} b_2 & c_2 \\ b_3 & c_3 \end{vmatrix} - a_2 \times \begin{vmatrix} b_1 & c_1 \\ b_3 & c_3 \end{vmatrix} + a_3 \times \begin{vmatrix} b_1 & c_1 \\ b_2 & c_2 \end{vmatrix}$$

$$= a_1(b_2c_3 - b_3c_2) - a_2(b_1c_3 - b_3c_1) + a_3(b_1c_2 - b_2c_1)$$

$$D_x = \begin{vmatrix} d_1 & b_1 & c_1 \\ d_2 & b_2 & c_2 \\ d_3 & b_3 & c_3 \end{vmatrix}$$

$$= d_1 \times \begin{vmatrix} b_2 & c_2 \\ b_3 & c_3 \end{vmatrix} - d_2 \times \begin{vmatrix} b_1 & c_1 \\ b_3 & c_3 \end{vmatrix} + d_3 \times \begin{vmatrix} b_1 & c_1 \\ b_2 & c_2 \end{vmatrix}$$

$$= d_1(b_2c_3 - b_3c_2) - d_2(b_1c_3 - b_3c_1) + d_3(b_1c_2 - b_2c_1)$$

$$D_y = \begin{vmatrix} a_1 & d_1 & c_1 \\ a_2 & d_2 & c_2 \\ a_3 & d_3 & c_3 \end{vmatrix}$$

$$= a_1 \times \begin{vmatrix} d_2 & c_2 \\ d_3 & c_3 \end{vmatrix} - a_2 \times \begin{vmatrix} d_1 & c_1 \\ d_3 & c_3 \end{vmatrix} + a_3 \times \begin{vmatrix} d_1 & c_1 \\ d_2 & c_2 \end{vmatrix}$$

$$= a_1(d_2c_3 - d_3c_2) - a_2(d_1c_3 - d_3c_1) + a_3(d_1c_2 - d_2c_1)$$

$$D_z = \begin{vmatrix} a_1 & b_1 & d_1 \\ a_2 & b_2 & d_2 \\ a_3 & b_3 & d_3 \end{vmatrix}$$

$$= a_1 \times \begin{vmatrix} b_2 & d_2 \\ b_3 & d_3 \end{vmatrix} - a_2 \times \begin{vmatrix} b_1 & d_1 \\ b_3 & d_3 \end{vmatrix} + a_3 \times \begin{vmatrix} b_1 & d_1 \\ b_2 & d_2 \end{vmatrix}$$

$$= a_1(b_2d_3 - b_3d_2) - a_2(b_1d_3 - b_3d_1) + a_3(b_1d_2 - b_2d_1)$$

3. Solve for x, y, and z using:

$$x = D_x/D$$

$$y = D_y/D$$

$$z = D_z/D$$

4. Check results by substituting x, y, and z into the original equations.

- **Example:** Solve the following three equations using method of determinants of a square matrix.

$$3x + 2y - 2z = 8$$

$$4x - 6y + 6z = 2$$

$$10x + 8y + 10z = -8$$

Create the four matrices using the coefficients a, b, c, and d in each equation. Then calculate the determinants of the four matrices.

$$D = \begin{vmatrix} 3 & 2 & -2 \\ 4 & -6 & 6 \\ 10 & 8 & 10 \end{vmatrix}$$

$$= 3 \times \begin{vmatrix} -6 & 6 \\ 8 & 10 \end{vmatrix} - 4 \times \begin{vmatrix} 2 & -2 \\ 8 & 10 \end{vmatrix} + 10 \times \begin{vmatrix} 2 & -2 \\ -6 & 6 \end{vmatrix}$$

$$= 3[(-6)(10) - (8)(6)] - 4[(2)(10) - (8)(-2)]$$
$$+ 10[(2)(6) - (-6)(-2)]$$

$$= 3(-60-48) - 4(20 - -16) + 10(12 - 12)$$

$$= 3(-108) - 4(36) + 10(0) = -324 - 144 = -468$$

$$D_x = \begin{vmatrix} 8 & 2 & -2 \\ 2 & -6 & 6 \\ -8 & 8 & 10 \end{vmatrix}$$

$$= 8 \times \begin{vmatrix} -6 & 6 \\ 8 & 10 \end{vmatrix} - 2 \times \begin{vmatrix} 2 & -2 \\ 8 & 10 \end{vmatrix} + -8 \times \begin{vmatrix} 2 & -2 \\ -6 & 6 \end{vmatrix}$$

$$= 8[(-6)(10) - (8)(6)] - 2[(2)(10) - (8)(-2)]$$
$$+ -8[(2)(6) - (-6)(-2)]$$

$$= 8(-60 - 48) - 2(20 - -16) + -8(12 - 12)$$

$$= 8(-108) - 2(36) + -8(0) = -864 - 72 = -936$$

$$D_y = \begin{vmatrix} 3 & 8 & -2 \\ 4 & 2 & 6 \\ 10 & -8 & 10 \end{vmatrix}$$

$$= 3 \times \begin{vmatrix} 2 & 6 \\ -8 & 10 \end{vmatrix} - 4 \times \begin{vmatrix} 8 & -2 \\ -8 & 10 \end{vmatrix} + 10 \times \begin{vmatrix} 8 & -2 \\ 2 & 6 \end{vmatrix}$$

$$= 3[(2)(10) - (-8)(6)] - 4[(8)(10) - (-8)(-2)]$$
$$+ 10[(8)(6) - (2)(-2)]$$

$$= 3(20 - -48) - 4(80 - 16) + 10(48 - -4)$$

$$= 3(68) - 4(64) + 10(52) = 204 - 256 + 520 = 468$$

$$D_z = \begin{vmatrix} 3 & 2 & 8 \\ 4 & -6 & 2 \\ 10 & 8 & -8 \end{vmatrix}$$

$$= 3 \times \begin{vmatrix} -6 & 2 \\ 8 & -8 \end{vmatrix} - 4 \times \begin{vmatrix} 2 & 8 \\ 8 & -8 \end{vmatrix} + 10 \times \begin{vmatrix} 2 & 8 \\ -6 & 2 \end{vmatrix}$$

$$= 3[(-6)(-8) - (8)(2)] - 4[(2)(-8) - (8)(8)]$$
$$+ 10[(2)(2) - (-6)(8)]$$

$$= 3(48 - 16) - 4(-16 - 64) + 10(4 - -48)$$

$$= 3(32) - 4(-80) + 10(52) = 96 + 320 + 520 = 936$$

Calculate x, y, and z.

$$x = D_x/D = -936/-468 = 2$$

$$y = D_y/D = 468/-468 = -1$$

$$z = D_z/D = 936/-468 = -2$$

Therefore, the solutions to:

$$3x + 2y - 2z = 8$$

$$4x - 6y + 6z = 2$$

$$10x + 8y + 10z = -8$$

are $x = 2$, $y = -1$, and $z = -2$.

Check the results by substituting into the original equations.

The first equation:

$$3x + 2y - 2z = 8$$

$$3(2) + 2(-1) - 2(-2) = 8$$

$$6 + -2 + 4 = 8$$

$$4 + 4 = 8$$

$$8 = 8$$

The second equation:

$$4x - 6y + 6z = 2$$

$$4(2) - 6(-1) + 6(-2) = 2$$

$$8 + 6 + -12 = 2$$

$$14 + -12 = 2$$

$$2 = 2$$

The third equation:

$$10x + 8y + 10z = -8$$

$$10(2) + 8(-1) + 10(-2) = -8$$

$$20 + -8 + -20 = -8$$

$$12 + -20 = -8$$

$$-8 = -8$$

- For information on solving systems of non-linear equations, refer to a college-level textbook.

Chapter 9

Working with Coordinate Systems and Graphing Equations

9.1 Introduction and Definitions

• Real numbers can be identified with points on a line and depicted on the *real number line.* If a number is represented by a point on the number line, the number is called the coordinate of that point. Pairs of real numbers that define a point on a plane can be depicted by identifying them with the two axes of a *two-dimensional coordinate system.* Points in three-dimensional space can be depicted by identifying them with the three axes of a *three-dimensional coordinate system.* Graphing on a coordinate system is often used to visualize quantitative data in a manner that will provide insight into trends, patterns, relationships, etc. Graphing an equation on a coordinate system provides a depiction of the slope (in the case of the linear equation) or the shape of a curve in the case of a non−linear equation. Geometrical figures defined by equations can be depicted on a coordinate system.

• Two-dimensional, planar *rectangular coordinate systems* consist of two axes, generally denoted x and y, for the horizontal and vertical axes that are at right angles to each other. A three-dimensional rectangular coordinate system consists of three axes, generally denoted x, y, and z, which are all three at right angles to each other. Rectangular coordinate systems are used in this chapter. However, it should be noted that there are other types of coordinate systems, such as *cylindrical* and *spherical.*

• In this section, the following are defined: Points on lines, points on two- and three-dimensional coordinate systems, perpendicular lines, parallel lines, and lines on coordinate systems.

- A *point* on the number line is represented in one dimension.

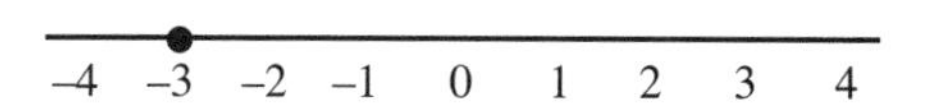

- A *point on planar coordinate system* with two intersecting axes that are perpendicular to each other is represented in two dimensions.

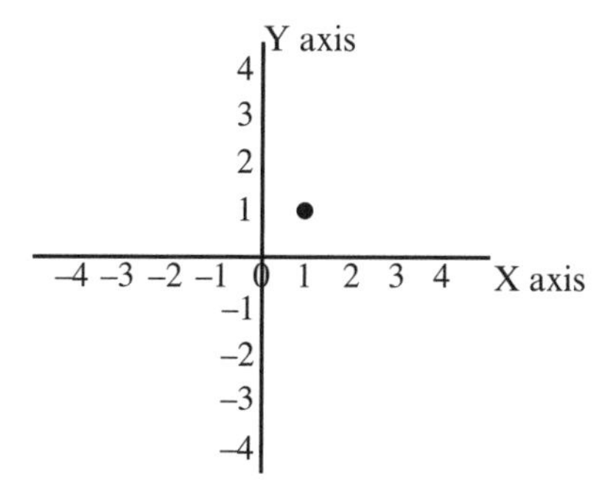

X is the horizontal axis. Y is the vertical axis. The axes intersect at zero on the X-axis and zero on the Y-axis. The point shown above corresponds to x = 1, y = 1.

- A *point on a spatial coordinate system* with three intersecting axes that are all perpendicular to each other is represented in three dimensions.

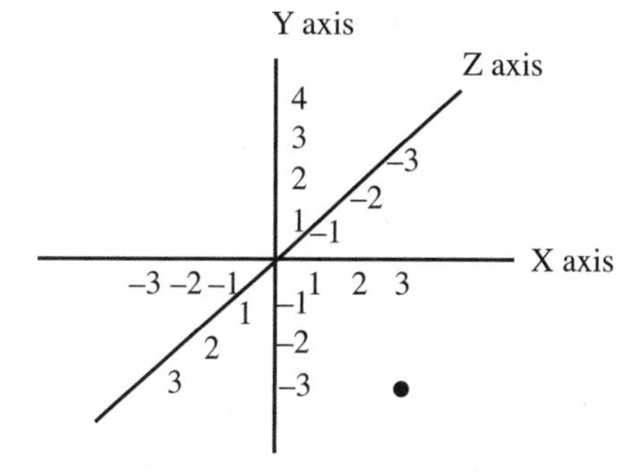

X is the horizontal axis. Y is the vertical axis. Z is the axis that comes out of the page in a horizontal direction. The point is represented in the 3-dimensional space. The axes intersect at zero on the X-axis, Y-axis, and Z-axis. The point corresponds to $x = 3$, $y = -3$, and $z = 3$. (The position of the point with respect to the Z-axis cannot be accurately visualized in two dimensions.)

- To define the position of a *point* on a line, the number the point corresponds to is identified.

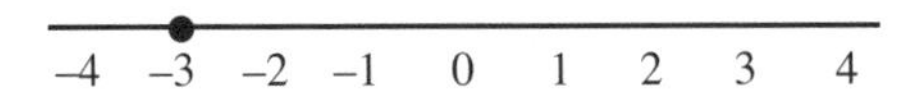

The point is at the -3 coordinate position.

- To define the position of a point on a two-axis planar coordinate system, the numbers on each axis that the point corresponds to are identified.

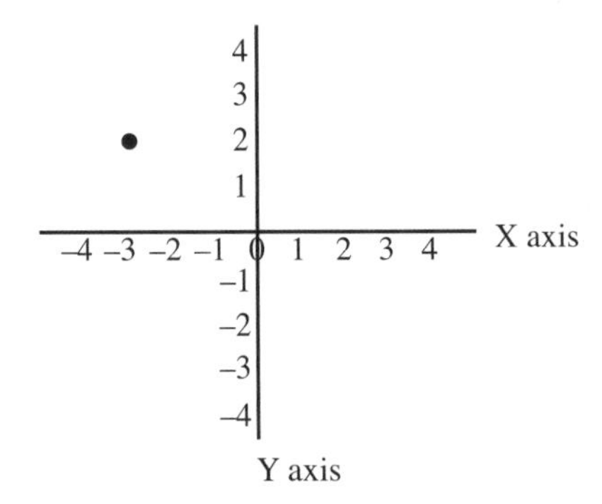

The point corresponds to -3 on the X-axis and to 2 on the Y-axis.

• The convention for writing numbers that a point corresponds to on each axis is, for the example above, (−3, 2):

(X-axis−number, Y-axis−number)

• To identify the point (x, y) on an X-Y coordinate system:

A positive x value is on the right of the Y-axis.

A negative x value is on the left of the Y-axis.

A positive y value is above the X-axis.

A negative y value is below the X-axis.

• To define the position of a point on a three-axis (three-dimensional) coordinate system the number on each axis that the point corresponds to is identified.

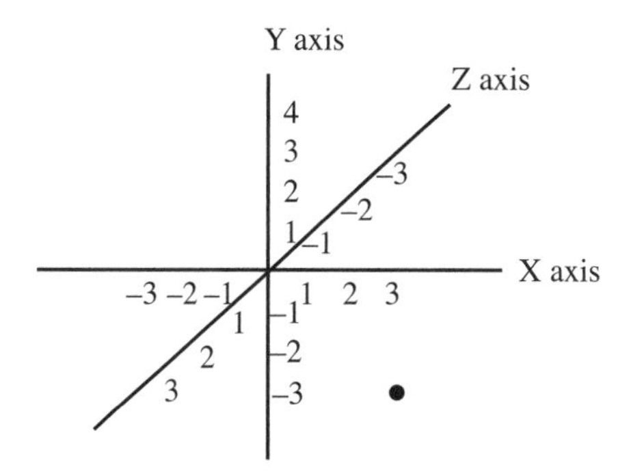

The point corresponds to +3 on the X-axis, −3 on the Y-axis, and +3 on the Z-axis, and can be identified by (x, y, z) = (3, −3, 3).

- A *line* can be defined by either a two-axis or a three-axis coordinate system, such that each point on the line corresponds to a position on each axis.

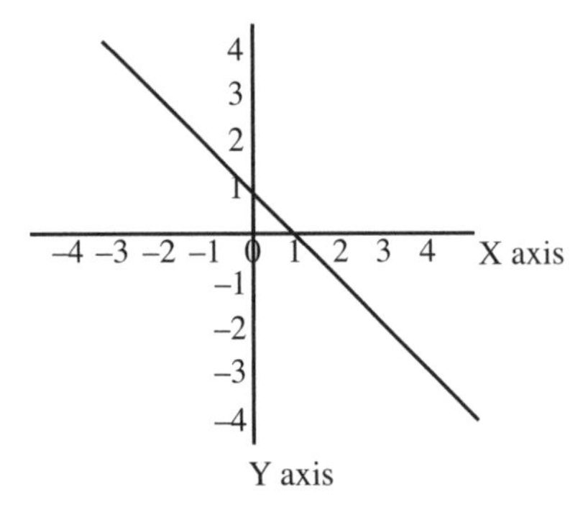

- *Perpendicular lines* intersect at right (90°) angles.

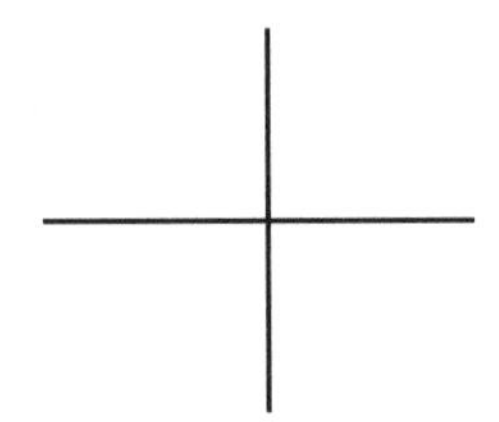

- *Parallel lines* do not intersect.

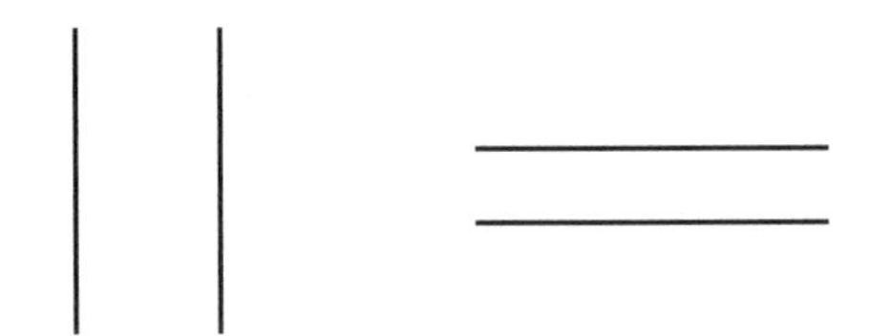

9.2 Graphing Linear Equations

• In this section, two methods for graphing linear equations and graphing inequalities are described.

• A *line* can be described by at least two points that correspond to positions on the X and Y axes or by a linear equation that contains variables x and y that correspond to points on the line.

• A *linear equation* contains only first-degree variables. Therefore, the variables cannot be in the form of exponents or radicals.

• If a linear equation is plotted, the result is a straight line. If a non-linear equation (having a variable with an exponent other than one) is plotted, the result is a curved line.

• The graph of the following equation is a straight line (if A and B are not both zero):

$$Ax + By = C$$

• One method used to graph an equation with the form $Ax + By = C$ is:

1. Simplify the equation by combining like terms.

2. Find the points where the line crosses (or intercepts) the X and Y axes by setting y equal to zero and solving for x, then setting x equal to zero and solving for y.

3. Graph the line by plotting the points at the intercepts and drawing the line through the points.

4. Check that each intercept equation was solved correctly by substituting a value into the original equation for x, solving for y, and making certain that this (x, y) point falls on the line.

- **Example:** Graph the equation:

$$4x - 6y - 8 = 8x + 2y + 8$$

Simplify the equation. Combine like terms.

$$4x - 8x - 6y - 2y = 8 + 8$$

$$-4x - 8y = 16$$

Set $x = 0$, and solve for y.

$$-4(0) - 8y = 16$$

$$-8y = 16$$

$$y = -16/8$$

$$y = -2$$

Therefore, the line intercepts the Y-axis at -2. Set $y = 0$, and solve for x.

$$-4x - 8(0) = 16$$

$$-4x = 16$$

$$x = -16/4$$

$$x = -4$$

Therefore, the line intercepts the X-axis at -4. Plot the intercept points and draw the line.

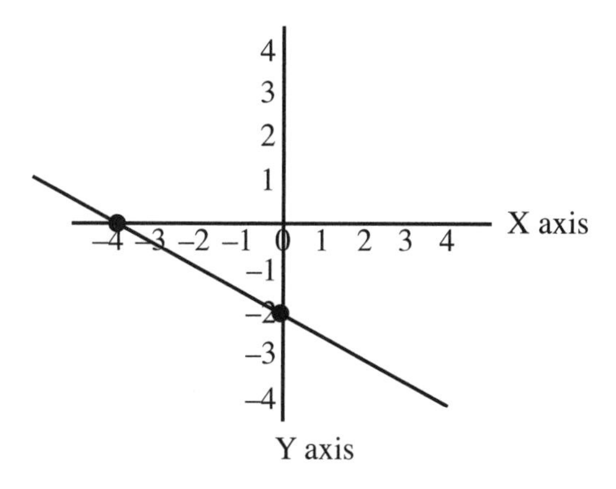

Check by substituting a value for x into the original (reduced) equation, $-4x - 8y = 16$. Solve for y, and verify that the new point falls on the plotted line.

Substitute $x = 1$ and solve for y.

$$-4(1) - 8y = 16$$

$$-4 - 8y = 16$$

$$-8y = 20$$

$$y = -20/8 = -2\ 1/2$$

The point $(1, -2\ 1/2)$ appears to fall on the line.

- An alternative method used to graph an equation with the form $Ax + By = C$ is:

1. Simplify the equation by combining like terms.

2. Choose at least two values for x and solve the equation for y.

3. Check that both values satisfy the equation by substituting them into the original equation.

4. Graph the line by plotting the points on the coordinate system and drawing the line through the points.

- **Example:** Graph the equation $y = x + 2$

The equation is already simplified.

Choose values for x and solve for y.

Choose $x = 1$, and solve for y.

$$y = 1 + 2$$

$$y = 3$$

The first resulting pair is (1, 3).

Check by substituting both x and y back into the original equation, $y = x + 2$.

$$3 = 1 + 2$$

$$3 = 3$$

Choose $x = -1$ and solve for y.

$$y = -1 + 2$$

$$y = 1$$

The second resulting pair is $(-1, 1)$.

Check by substituting both x and y back into the original equation, $y = x + 2$.

$$1 = -1 + 2$$

$$1 = 1$$

Choose $x = 0$, and solve for y.

$$y = 0 + 2$$

$$y = 2$$

The third resulting pair is $(0, 2)$.

Check by substituting both x and y back into the original equation, $y = x + 2$.

$$2 = 0 + 2$$

$$2 = 2$$

Choose $x = 2$, and solve for y.

$$y = 2 + 2$$

$$y = 4$$

The fourth resulting pair is $(2, 4)$.

Check by substituting both x and y back into the original equation, $y = x + 2$.

$$4 = 2 + 2$$

$$4 = 4$$

Choose $x = -2$, and solve for y.

$$y = -2 + 2$$

$$y = 0$$

The fifth resulting pair is $(-2, 0)$.

Check by substituting both x and y back into the original equation, $y = x + 2$.

$$0 = -2 + 2$$

$$0 = 0$$

Plot points $(1, 3)$, $(-1, 1)$, $(0, 2)$, $(2, 4)$, $(-2, 0)$, and draw the line.

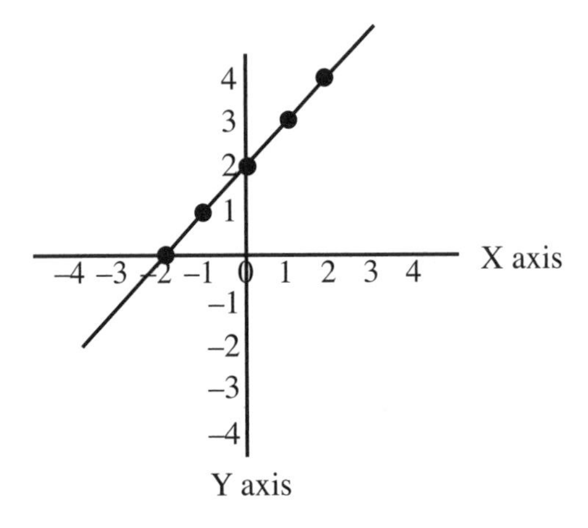

• To *graph an inequality*, first graph the equality as described above. The solution to the inequality encompasses all of the points on one or the other side of the plotted points.

If the inequality is $>$ or $<$, the solution does not include the line drawn through the points. In this case draw a dotted line through the points and use shading on one side of the dotted line to depict the solution set.

If the inequality is $\leq$ or $\geq$, the solution does include the line drawn through the points. In this case draw a solid line through the points and use shading on one side of the dotted line to depict the solution set.

To determine which side of the line comprises the solution, select a point on one side and substitute it into the inequality. If it satisfies the inequality, the solution set consists of all the points on that side of the line. If the point does not satisfy the inequality, the solution set consists of all the points on the other side of the line.

9.3 Slope of a Line

• Determining the slope of a line, equations for lines and the equation for the slope of a line are described in this section.

• The *slope* of a line on a coordinate system can be determined by choosing a segment of the line between two convenient points and calculating the change in the vertical y-direction divided by the change in the horizontal x-direction.

The change in the y-direction is determined by subtracting the corresponding Y-axis value of the *first* y-point from the corresponding Y-axis value of the *second* y-point. The change in the x-direction is determined by subtracting the corresponding X-axis value of the *first* x-point from the corresponding X-axis value of the *second* x-point.

- The slope is often referred to as *m*.

- The *equation for the slope of a line* is:

$$m = \frac{y_2 - y_1}{x_2 - x_1}$$

- To find the slope of the line in the graph below:

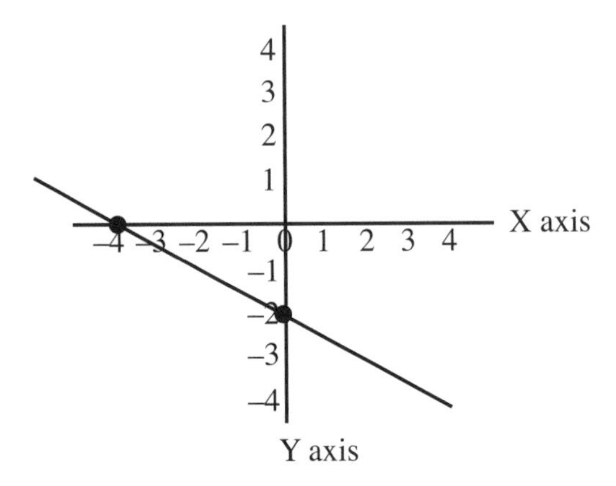

Choose the points shown, (−4, 0) and (0, −2), and using the equation for a line:

$$m = \frac{y_2 - y_1}{x_2 - x_1}$$

Substitute the values for x_1, y_1, x_2, and y_2.

$$m = \frac{-2 - 0}{0 - -4}$$

Reduce the equation.

$$m = \frac{-2}{4} = -\frac{1}{2}$$

Therefore, the slope of the line is $-1/2$.

- The relationship between slope and the line is as follows:

If the slope is a positive number, the line is inclined upward.

If the slope is a negative number, the line is inclined downward.

If the slope is zero, the line is horizontal.

The slope of a vertical line is undefined because it would require a denominator of zero.

- The equation for the slope of a line, $m = \frac{y_2 - y_1}{x_2 - x_1}$

can be equivalently written, $y_2 - y_1 = m(x_2 - x_1)$.

This equation can be used to find the equation of a particular line if the coordinates of one point on the line and the slope are known.

- For example, if one point and the slope of a line are indentified, the equation describing the line is:

$$y - y_0 = m(x - x_0)$$

Where (x_0, y_0) represent the known point.

If the known point, (x_0, y_0), is on the Y-axis at the position where the line intercepts the Y-axis, the known point is $(0, y_0)$. (x is zero.)

In this case, the equation can be written:

$$y - y_0 = m(x - 0)$$

$$y - y_0 = mx$$

$$y = mx + y_0$$

This equation is commonly used to find the slope and y-intercept point when the equation of the line is given. It is generally written with b representing y_0.

$$y = mx + b$$

In summary the equations used to describe straight lines are as follows:

- The following equation is used to graph a linear equation (by choosing values for x and solving for y, then plotting the pairs of (x, y) values), and to solve for the unknown variables, x and y, in systems of equations:

$$Ax + By = C$$

- The following equation is used to *find the equation of a line when the slope of the line and one point on the line are identified:*

$$y - y_0 = m(x - x_0)$$

For example, if the slope of a line is -3 and one point on the line is $(-1, 4)$, the equation for the line is:

$$y - 4 = -3(x + 1)$$

- The following equation is used to *find the slope (m) and y-intercept (b) of a line when the equation is known*:

$$y = mx + b$$

For example, if the equation is $2x + 6y = 12$, or equivalently, arranging in the form of $y = mx + b$:

$$6y = -2x + 12$$

$$y = (-2x + 12)/6$$

$$y = (-1/3)x + 2$$

Comparing with $y = mx + b$,

$$m = -1/3 \text{ and } b = 2$$

- The following equation is used to *find the equation of a line when two points on the line are identified:*

$$\frac{y - y_1}{x - x_1} = \frac{y_2 - y_1}{x_2 - x_1}$$

- The following equation is used to *find the equation of a line when two intercept points (a, 0) on the X-axis and (0, b) on the Y-axis are identified*:

$$\frac{x}{a} = \frac{y}{b} = 1$$

9.4 Graphs of the Equations for the Parabola

- Graphs of non-linear equations can form parabolas.

- The equation for a *parabola* is in the form:

  $y = ax^2 + bx + c$ for a parabola with a vertical axis.

  $x = ay^2 + by + c$ for a parabola with a horizontal axis.

Notice the similarity between these equations and the quadratic equation. The graph of a quadratic equation has the form of a parabola.

- The following are *vertical-axis parabolas*:

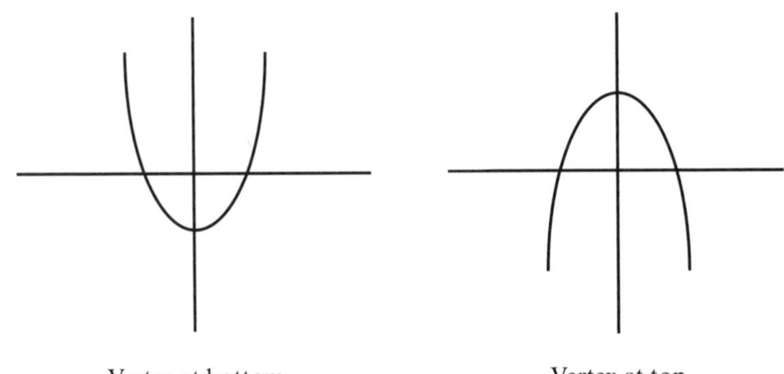

Vertex at bottom Vertex at top

- In the vertical form of the equation, $y = ax^2 + bx + c$, if a is positive, the parabola is open at the top with the vertex at the bottom. Conversely, if a is negative, the parabola is open at the bottom with the vertex at the top.

- The following is a *horizontal-axis parabola* with its vertex at the right:

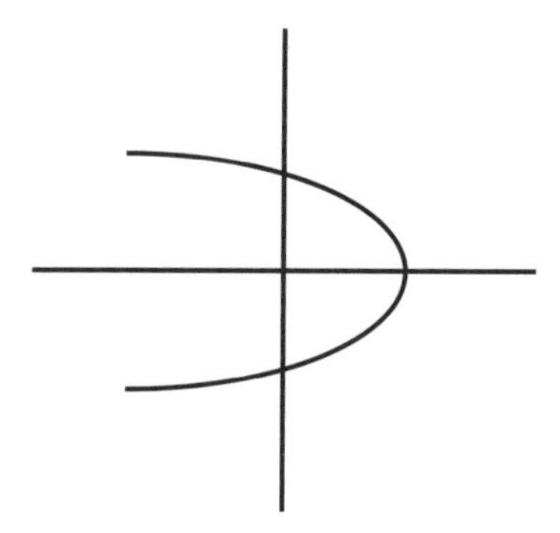

- A line called the *axis of symmetry* can be drawn through the center of a parabola to divide it in half.

- The equation for the axis of symmetry in a vertical parabola is:

$$x_v = -b/2a$$

If the solution to this equation is $x_v = 2$, then the axis of symmetry can be drawn as a vertical line through the point at 2 on the X-axis.

If $x_v = 2$ is substituted for x into the following equation:

$$y = ax^2 + bx + c,$$

and the equation is solved for the corresponding y or y_v value, the resulting (x_v, y_v) pair represents the point at the vertex of the parabola. (See the following section.)

9.5 Graphing Quadratic Equations

- The graph of a *quadratic equation* is a parabola. The two methods described below can be used for graphing quadratic equations.

- One method of graphing quadratic equations involves finding the vertex point $x_v = -b/2a$, substituting x_v into the quadratic equation form $y_v = ax_v^2 + bx_v + c$, and solving for y_v resulting in the vertex point (x_v, y_v). Then, choose other values for x on both sides of x_v and solve for their corresponding y values using the original equation. Finally, plot the points and sketch the parabola.

- An alternative method for graphing quadratic equations involves choosing values for x and solving the quadratic equation for the corresponding y values. Then, plot the points and sketch the parabola.

- It is faster to use the first method and to find the vertex point first, because using the second method may require identifying many points before finding both legs of the parabola.

- **Example:** Graph the quadratic equation:

  $2x^2 - 4x + 4 = 2y$

Solve for y. Divide both sides by 2.

$y = (1/2)(2x^2 - 4x + 4)$

$y = x^2 - 2x + 2$

Find the vertex point (x_v, y_v) using $x_v = -b/2a$, and write the quadratic equation in standard form.

$y = ax^2 + bx + c.$

$y = x^2 - 2x + 2$

In this example, $b = -2$ and $a = 1$. Solving for x_v,

$x_v = -b/2a$

$x_v = -(-2)/2(1)$

$x_v = 2/2 = 1$

Solve for y_v using the original equation (written in standard form).

$y = x^2 - 2x + 2$

$y = (1)^2 - 2(1) + 2$

$y = 1 - 2 + 2$

$y = -1 + 2$

$y = 1$

The vertex point is (1, 1).

Choose other values for x on both sides of x_v and solve the original equation for y.

Choose x = 2 and solve for y.

$$y = x^2 - 2x + 2$$

$$y = 2^2 - 2(2) + 2$$

$$y = 4 - 4 + 2$$

$$y = 2$$

The pair is (2, 2).

Choose x = 0 and solve for y.

$$y = x^2 - 2x + 2$$

$$y = 0^2 - 2(0) + 2$$

$$y = 0 - 0 + 2$$

$$y = 2$$

The pair is (0, 2).

Choose x = −1 and solve for y.

$$y = x^2 - 2x + 2$$

$$y = (-1)^2 - 2(-1) + 2$$

$$y = 1 - -2 + 2$$

$$y = 3 + 2$$

$$y = 5$$

The pair is (−1, 5).

Choose x = 3 and solve for y.

$$y = x^2 - 2x + 2$$

$$y = 3^2 - 2(3) + 2$$

$$y = 9 - 6 + 2$$

$$y = 5$$

The pair is (3, 5).

Plot points (1, 1), (2, 2), (0, 2), (−1, 5), (3, 5), and sketch the parabola.

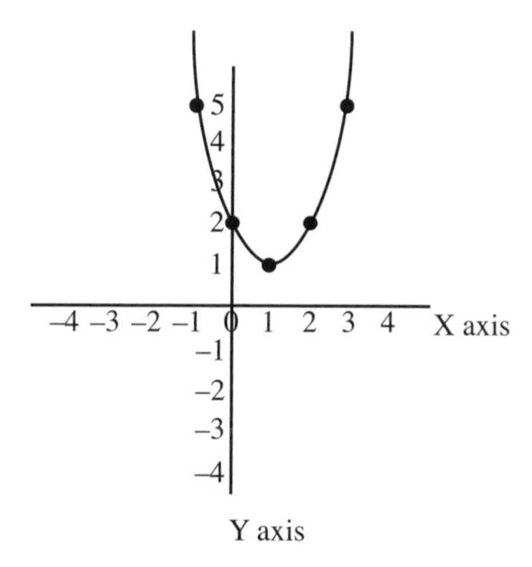

9.6 Using Graphing to Solve Quadratic Equations

- The standard form of a quadratic equation is:

$$ax^2 + bx + c = 0$$

For graphing, replace 0 with y.

$$y = ax^2 + bx + c$$

- To solve a quadratic equation graphically:

1. Write the equation in the standard form by combining like terms and replace zero with y.

2. Graph the parabola that corresponds to the quadratic equation by identifying the vertex point $x_v = -b/2a$ and solving for y_v.

3. Choose other x values near the vertex point and solve for corresponding y values using the quadratic equation.

4. Solutions for x (called the roots of the equation because of the x^2 term) are determined by estimating the two points where the parabola crosses the X-axis (at $y = 0$).

- **Example:** Solve the quadratic equation, $2x^2 = 2x + 2$, by graphing.

First rearrange.

$$0 = 2x^2 - 2x - 2$$

Write in standard form, $y = ax^2 + bx + c$, and replace zero with y.

$$y = 2x^2 - 2x - 2$$

Graph the parabola by determining the vertex point and other surrounding points.

$$x_v = -b/2a$$

Comparing with the standard form equation.

$$a = 2 \text{ and } b = -2$$

Substituting for a and b:

$$x_v = -(-2)/2(2)$$

$$x_v = 2/4 = 1/2$$

Determine y_v.

$$y_v = 2x^2 - 2x - 2$$

Substitute x_v for x.

$$y_v = 2(1/2)^2 - 2(1/2) - 2$$

$$y_v = 2(1/4) - 2/2 - 2$$

$$y_v = 1/2 - 1 - 2$$

Using a common denominator of 2, the equation becomes:

$$y_v = 1/2 - 2/2 - 4/2$$

$$y_v = \frac{1-2-4}{2}$$

$$y_v = -5/2$$

Therefore, the vertex point is (1/2, −5/2).

Choose x values near the vertex point and solve for the corresponding y values.

Choose x = 1 and solve for y.

$$y = 2x^2 - 2x - 2$$

$$y = 2(1)^2 - 2(1) - 2$$

$$y = 2 - 2 - 2$$

$$y = -2$$

The pair is (1, −2).

Choose x = −1 and solve for y.

$$y = 2x^2 - 2x - 2$$

$$y = 2(-1)^2 - 2(-1) - 2$$

$$y = 2 - -2 - 2$$

$$y = 2$$

The pair is (−1, 2).

Choose $x = 0$ and solve for y.

$$y = 2x^2 - 2x - 2$$

$$y = 2(0)^2 - 2(0) - 2$$

$$y = 0 - 0 - 2$$

$$y = -2$$

The pair is $(0, -2)$.

Choose $x = 2$ and solve for y.

$$y = 2x^2 - 2x - 2$$

$$y = 2(2)^2 - 2(2) - 2$$

$$y = 8 - 4 - 2$$

$$y = 2$$

The pair is $(2, 2)$.

Plot points $(1/2, -5/2)$, $(1, -2)$, $(-1, 2)$, $(0, -2)$, $(2, 2)$, and sketch the parabola.

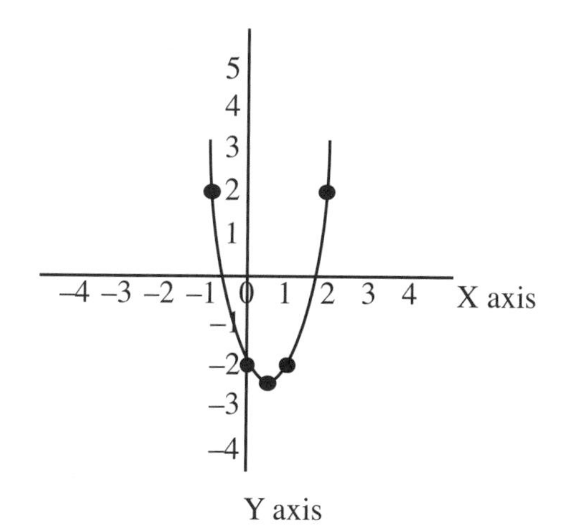

Determine the solutions to x by estimating the two points where the parabola crosses the X-axis (at $y = 0$).

$$x = +1.62$$

$$x = -0.62$$

Check by substituting each point into the original equation, $0 = 2x^2 - 2x - 2$.

For $x = +1.62$,

$$0 = 2(1.62)^2 - 2(1.62) - 2$$

$$0 = 2(2.62) - 3.24 - 2$$

$$0 = 5.24 - 3.24 - 2$$

$$0 = 2 - 2$$

$$0 = 0$$

For $x = -0.62$,

$$0 = 2(-0.62)^2 - 2(-0.62) - 2$$

$$0 = 2(0.38) + 1.24 - 2$$

$$0 = 0.76 + 1.24 - 2$$

$$0 = 2 - 2$$

$$0 = 0$$

Therefore, the estimated solutions to $0 = 2x^2 - 2x - 2$ are $x = 1.62$ and $x = -0.62$.

9.7 Using Graphing to Solve Two Linear Equations with Two Unknown Variables

- To solve for two unknown variables in two linear equations by graphing, the following procedure can be used:

1. Let one unknown variable be represented by x and the other by y.

2. Plot each equation separately as a straight line by setting $x = 0$ and solving for y, then setting $y = 0$ and solving for x. (Or by choosing values for x and solving for corresponding y values.)

3. Draw the two lines.

4. Solutions for x and y correspond to the point where the two lines intersect each other.

5. Check the solutions by substituting them into each of the original equations.

- **Example:** Solve the two equations for the two unknown variables by graphing.

$$2x + 2y - 8 = 0$$

$$4x - 8y - 8 = 0$$

Reduce each equation by dividing by 2 in the first equation and by 4 in the second.

$$x + y - 4 = 0$$

$$x - 2y - 2 = 0$$

In the first equation, $x + y - 4 = 0$,

Let $x = 0$, and solve for y.

$$0 + y - 4 = 0$$

$$y = 4$$

The point where the first equation crosses the Y-axis is (0, 4).

Let $y = 0$, and solve for x.

$$x + 0 - 4 = 0$$

$$x = 4$$

The point where the first equation crosses the X-axis is (4, 0).

The points for the first equation are (0, 4) and (4, 0).

In the second equation, $x - 2y - 2= 0$,

Let $x = 0$, and solve for y.

$$(0) - 2y - 2 = 0$$

$$-2y = 2$$

$$y = -1$$

The point where the second equation crosses the Y-axis is $(0, -1)$.

Let $y = 0$, and solve for x.

$$x - 2(0) - 2 = 0$$

$$x - 2 = 0$$

$$x = 2$$

The point where the second equation crosses the X-axis is $(2, 0)$.

The points for the second equation are $(0, -1)$ and $(2, 0)$.

Graph the two equations and find the solution by identifying the point where they intersect each other.

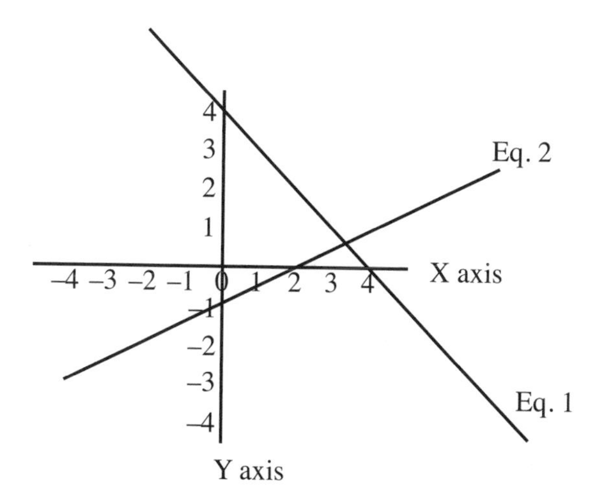

The intersection point of the two lines is (3 1/3, 2/3).

Therefore, the solutions to $2x + 2y - 8 = 0$ and $4x - 8y - 8 = 0$ are:

$$x = 3\ 1/3 = 10/3 \text{ and } y = 2/3$$

Check the solutions by substituting into each of the original equations.

The first equation:

$$x + y - 4 = 0$$

$$10/3 + 2/3 = 4$$

$$12/3 = 4$$

$$4 = 4$$

The second equation:

$$x - 2y - 2 = 0$$

$$10/3 - 2(2/3) = 2$$

$$\frac{10}{3} - \frac{4}{3} = 2$$

$$\frac{6}{3} = 2$$

$$2 = 2$$

• Systems of two linear equations are categorized as follows:

A system of two equations is called "consistent and independent" if there is only one solution for each variable. In this case the two lines described by the equations intersect at one point.

A system of two equations is called "inconsistent" if there are no solutions. In this case the two lines described by the equations are parallel and do not intersect at any point.

A system of two equations is called "consistent and dependent" if there are unlimited solutions for each variable. In this case the two lines described by the equations are the same line.

9.8 Examples of Other Equation Forms That Graph to Shapes on a Coordinate System

• In this section, equations and graphs of ellipses, circles, and hyperbolas are presented.

• *Ellipses* are oval shapes. The following two equations represent an ellipse with its origin at the (0, 0) point of a coordinate system:

$$Ax^2 + By^2 = C \quad \text{and} \quad \frac{x^2}{a^2} + \frac{y^2}{b^2} = 1$$

Where $A \neq B$ and A, B, and C are positive numbers.

An ellipse with its origin at some point (p, q) is represented by the equation

$$\frac{(x-p)^2}{a^2} + \frac{(y-q)^2}{b^2} = 1$$

- To plot an equation of an ellipse, choose values for x and solve for the corresponding y values. Or set x = 0 and solve for the corresponding y value and set y = 0 and solve for the corresponding x value.

- The following is an example of an ellipse with points (2, 0), (−2, 0), (0, 3), and (0, −3):

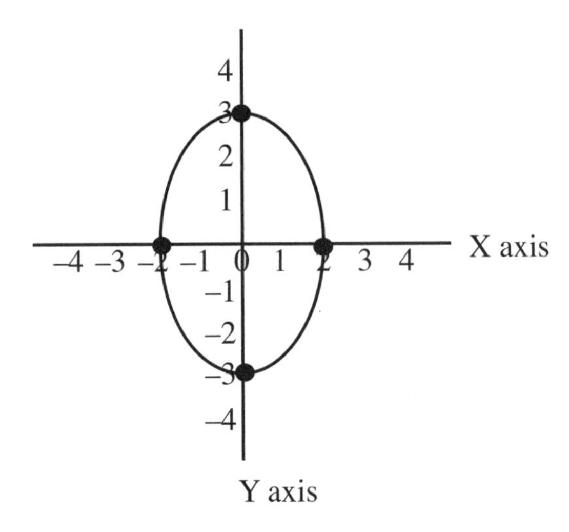

- The following equations represent a *circle* with its origin at the (0, 0) point of the coordinate system:

$$x^2 + y^2 = C \text{ or } Ax^2 + By^2 = C$$

Where A = B and A, B, and C are positive numbers.

$x^2 + y^2 = r^2$ Where r = radius.

A circle with its origin at some point (p, q) is represented by the equation:

$$(x - p)^2 + (y - q)^2 = r^2$$

- To plot an equation of a circle, choose values for x and solve for the corresponding y values. Alternatively, set x = 0 and solve for the corresponding y value and/or set y = 0 and solve for the corresponding x value.

If the circle has its origin at (0, 0), it is possible to choose x values in one quadrant of the coordinate system and use symmetry to complete the circle.

- The following is an example of a circle with points (2, 0), (−2, 0), (0, 2), and (0, −2):

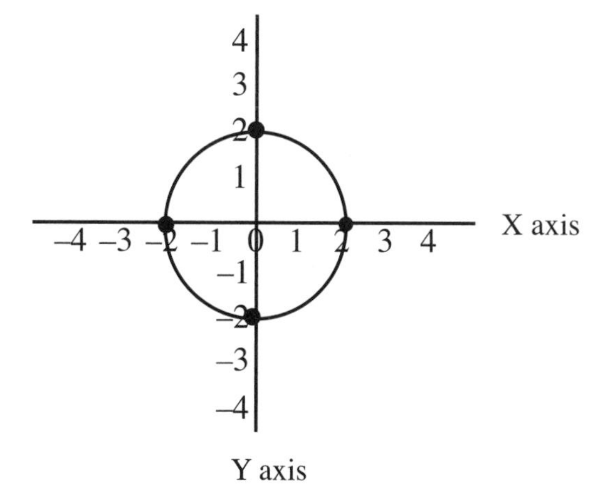

• The following equations represent a *hyperbola* with its origin at the (0, 0) point of the coordinate system:

$$-Ax^2 + By^2 = C \text{ or } Ax^2 + -By^2 = C$$

$$\frac{x^2}{a^2} - \frac{y^2}{b^2} = 1$$

A hyperbola with its origin at some point (p, q) is represented by the equation:

$$\frac{(x-p)^2}{a^2} - \frac{(y-q)^2}{b^2} = 1$$

• To plot an equation of a hyperbola, choose values for x and solve for the corresponding y values.

• The following is an example of a hyperbola:

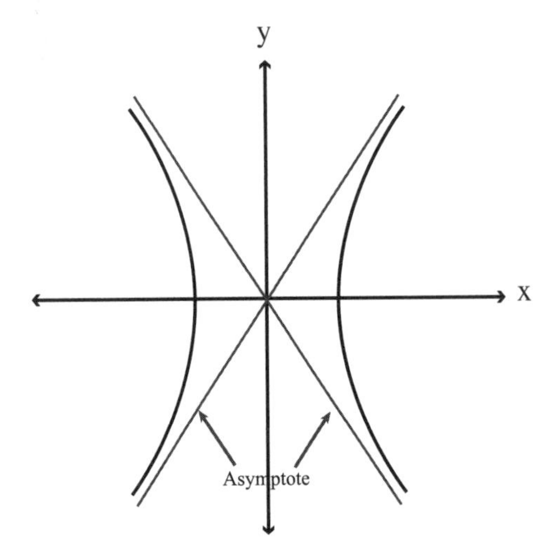

This hyperbola is along the X axis. A hyperbola can also be depicted along the Y axis. A hyperbola is symmetric with respect to the X and/or Y axes.

The heavy diagonal lines are called *asymptotes*. The equations for the asymptotes are:

$$-Ax^2 + By^2 = 0 \text{ or } Ax^2 + -By^2 = 0$$

Another equation for a hyperbola is:

$$xy = k$$

If k is positive, the hyperbola will graph in the upper right and lower left quadrants. If k is negative, the hyperbola will graph in the upper left and lower right quadrants.

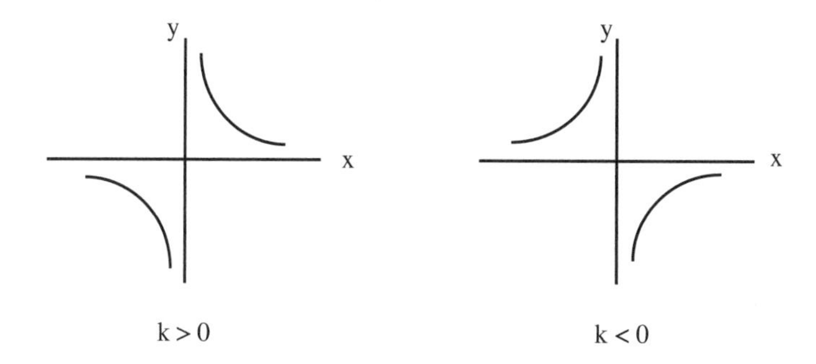

Index

D

E

F

G

H

I

L

M

N

O

P

Q

R

S

T

U

V

W

X

Y

Z

Notes

Notes

Notes

Notes